Antimicrobial Substances in Plants: Discovery of New Compounds, Properties, Food and Agriculture Applications, and Sustainable Recovery

Antimicrobial Substances in Plants: Discovery of New Compounds, Properties, Food and Agriculture Applications, and Sustainable Recovery

Editors

Daniela Gwiazdowska
Krzysztof Juś
Katarzyna Marchwińska

MDPI • Basel • Beijing • Wuhan • Barcelona • Belgrade • Manchester • Tokyo • Cluj • Tianjin

Editors

Daniela Gwiazdowska
Department of Natural Science and Quality Assurance,
Poznań University of Economics and Business,
Poznan, WP, Poland

Krzysztof Juś
Department of Natural Science and Quality Assurance,
Poznań University of Economics and Business,
Poznan, WP, Poland

Katarzyna Marchwińska
Department of Natural Science and Quality Assurance,
Poznań University of Economics and Business,
Poznan, WP, Poland

Editorial Office
MDPI
St. Alban-Anlage 66
4052 Basel, Switzerland

This is a reprint of articles from the Special Issue published online in the open access journal *Applied Sciences* (ISSN 2076-3417) (available at: https://www.mdpi.com/journal/applsci/special_issues/antimicrobial_substances_in_plants).

For citation purposes, cite each article independently as indicated on the article page online and as indicated below:

LastName, A.A.; LastName, B.B.; LastName, C.C. Article Title. *Journal Name* **Year**, *Volume Number*, Page Range.

ISBN 978-3-0365-4425-0 (Hbk)
ISBN 978-3-0365-4426-7 (PDF)

Contents

About the Editors

Daniela Gwiazdowska

Daniela Gwiazdowska is a Professor at Poznań University of Economics and Business, ORCID: 0000-0002-0972-6225. The area of scientific research includes: (i) design and quality assessment of fermented food products; (ii) biological control in plant protection: (iii) antimicrobial and anti-mycotoxigenic activity of essential oils and plant extracts; (iv) decontamination of *Fusarium* mycotoxins with microorganisms and essential oils; (v) application of molecular biology methods for the detection and identification of microorganisms; (vi) fermentation as waste management strategy.

Krzysztof Juś

Krzysztof Juś is an Assistant Professor at Poznań University of Economics and Business, ORCID: 0000-0001-8113-5481. The area of scientific research includes: (i) microbial quality and safety of products; (ii) zero-waste approach in new product development; (iii) biological methods of *Fusarium* fungi and mycotoxins' decontamination; (iv) design and quality assessment of natural disinfectants and preservatives; (v) application of molecular biology methods for the detection and identification of microorganisms.

Katarzyna Marchwińska

Katarzyna Marchwińska is an Assistant Professor at Poznań University of Economics and Business, ORCID: 0000-0001-9546-5334. The area of scientific research includes: (i) design and quality assessment and functional properties of bioactive food, probiotic preparations and feed additives; (ii) the use of microorganisms and bioactive substances to improve the quality and preservation of food and feed; (iii) application of new microorganisms and natural compounds in pathogen control techniques and their toxic metabolites in the food chain; (iv) search, design and evaluation of the effectiveness of new compounds ensuring adequate hygiene of production and packaging; (v) application of molecular biology methods for the detection and identification of microorganisms.

Editorial

Special Issue "Antimicrobial Substances in Plants: Discovery of New Compounds, Properties, Food and Agriculture Applications, and Sustainable Recovery"

Daniela Gwiazdowska *, Katarzyna Marchwińska and Krzysztof Juś

Department of Natural Science and Quality Assurance, Institute of Quality Science, Poznań University of Economics and Business, Niepodległości 10, 61-875 Poznań, Poland;
katarzyna.marchwinska@ue.poznan.pl (K.M.); krzysztof.jus@ue.poznan.pl (K.J.)
* Correspondence: daniela.gwiazdowska@ue.poznan.pl; Tel.: +48-61-856-9536

Citation: Gwiazdowska, D.; Marchwińska, K.; Juś, K. Special Issue "Antimicrobial Substances in Plants: Discovery of New Compounds, Properties, Food and Agriculture Applications, and Sustainable Recovery". *Appl. Sci.* **2022**, *12*, 5021. https://doi.org/10.3390/app12105021

Received: 26 April 2022
Accepted: 10 May 2022
Published: 16 May 2022

Microbial contamination of agriculture and food commodities may cause significant losses, with economic, social and environmental consequences. Therefore, the search for new, promising substances that demonstrate antagonism towards different microorganisms has been observed in recent years. Plants are a valuable source of different bioactive compounds that exhibit antimicrobial activity. These substances usually play a key role as a defense factor against different microorganisms and predators, as well as acting as growth regulators. Plant antimicrobials are varied in their chemical nature and include different compounds, such as polyphenols, terpenoids and essential oils, alkaloids, lectins and polypeptides. Due to the high diversity of their structure and properties, these substances may be applied against bacteria and fungi in the whole food chain, including pathogens and spoilage microorganisms. Taking into account the increased interest in natural antimicrobials, plant metabolites seem to be an important alternative for chemical pesticides in plant protection, as well as for preservatives in food or food packaging. Therefore, research on the discovery of new substances and their antimicrobial activity against bacteria and toxigenic fungi occurring in food and food processing, as well as those responsible for plant infections during their growth, is expanding the current knowledge about plant metabolites. Considering that the majority of research concerns planktonic cells, the activity of plant antimicrobials should be equally important in relation to biofilms formed by pathogenic and spoilage microorganisms. Moreover, the implementation of new technologies, as well as the sustainable recovery of antimicrobial substances from waste materials, is a subject of concern.

The contributed articles belong to the following two groups: (i) concentrated on the biological activity of the plant-derived extracts and oils as well as (ii) novel extraction techniques for obtaining plant derived-extracts.

Plant extracts accommodate a wide diversity of secondary metabolites, which are useful as alternative strategies to control infectious diseases, but are also increasingly used as preservatives in the food industry, in pharmaceuticals and cosmetics and as natural fungicides in agriculture. Nowadays, studies concerning the effect of plant extracts are focusing on the qualitative and quantitative state of the bioactive compounds (e.g., phenolic acids, flavonoids) and their interactions with different microorganisms within the food, cosmetic and agricultural products. A few papers are focused on the study of the phenolic content and antioxidant capacities of, subjected to their studies, plant extracts, indicating the relation of these values on the biological properties of extracts, including antibacterial, antifungal as well as cytotoxic activity.

Ziarno et al. (2021) [1] tested the viability of lactic acid bacteria (LAB) during the fermentation and passage of milk products, enriched with different plant extracts, through the in vitro gastrointestinal digestive tract. In the studies of *Valeriana officinalis* L., *Salvia officinalis* L., *Matricaria chamomilla* L., *Cistus* L., *Tilia* L., *Plantago lanceolata* L. and *Althaea* L., plant extracts were added at the beginning of the controlled milk fermentation process in order to obtain information about their interaction with the starter cultures. The results showed that small amounts of the extracts did not influence the fermentation processes; therefore, yogurts enriched with such extracts enable the introduction of new functional food products to the market [1].

In turn, in the research by Gwiazdowska et al. (2022) [2], fungistatic activity of cinnamon bark (*Cinnamomum zeylanicum*), verbena leaves and flowers (*Thymus hiemalis*) and palmarosa leaves (*Cymbopogon martini*) essential oils (EOs) against *F. graminearum* in corn and wheat grain was investigated. All the tested EOs effectively inhibited the growth of fungi in concentrations of 5% and 10%. Cinnamon and verbena EOs also effectively reduced the ergosterol (ERG) content in both grains at the concentration of 1%, while at the 0.1% EO concentration, the reduction in the ERG amount depended on the EO type, as well as on the grain. Moreover, the effect of EOs on the reduction in biosynthesis of the two main *Fusarium* mycotoxins deoxynivalenol and zearalenone has been shown. The degree of zearalenone (ZEA) reduction was consistent with the inhibition of ERG biosynthesis; however, the reduction in deoxynivalenol (DON) was not consistent with this parameter [2].

Nguenha et al. (2021) [3] described the use of plant-derived substances to reduce the growth and mycotoxin production by filamentous fungi in maize grains. The authors evaluated the effect of the combination of photosensitization and curcumin to inactivate *Aspergillus flavus* spores and minimize aflatoxin B1 (AFB1) production, in vitro, in maize grain. In this study, the maintenance of carotenoid content in three maize varieties after photosensitization was also evaluated. Photosensitization of maize resulted in a complete viability reduction in *A. flavus* and, therefore, the inhibition of AFB1 production, but did not affect the carotenoid content. On the other hand, no significant effect was observed using either light or curcumin. The obtained results indicate that photosensitization may be a successful green preservation method for AFB1 decontamination of the maize, without any negative effect on carotenoid content [3].

The studies performed by Petropoulos et al. (2021) [4] concerned linseed, purslane, luffa, and pumpkin seed oils' antimicrobial and cytotoxic activities, along with the fatty acid compositions. The obtained results showed the highest antibacterial activity of linseed oil and pumpkin seed oil towards most of the tested bacteria, with the most promising results towards *Enterobacter cloacae* and *Escherichia coli*; the MIC and MBC values of these seed oils were similar to the positive controls used. All the tested seed oils were determined with antifungal properties, obtaining results more effective than the positive controls in the case of *Aspergillus versicolor*, *Aspergillus niger* and *Penicillium verrucosum* var. *cyclopium*. Regarding the cytotoxic properties, luffa seed oil was the most effective against the tested HeLa cancer cell lines (cervical carcinoma) and NCI-H460 (non-small cell lung cancer). Petropoulos et al. 2021 [4] proved that selected vegetable seed oils pose promising antimicrobial and cytotoxic properties, which might be related to the fatty acid composition of the tested oils.

Plants, due to the large biological and structural diversity of their components, constitute unique and renewable sources for the discovery of new antibacterial, and antifungal substances. Recent insights regarding the possibilities of fruit and vegetable waste-derived antimicrobial substances, which contain a wide variety of secondary metabolites, are proving their usefulness as alternative strategies for food, cosmetic and agriculture applications.

The extraction techniques of the plant-derived bioactive compounds have a significant impact on the quality of the extracts and their chemical composition. Therefore, new sustainable extraction methods have emerged in recent years that provide faster and more efficient transfer of solutes from the sample to the solvent. Using different optimized extraction approaches may enable plant extracts with more defined, stronger antimicrobial activity to be obtained, compared to the ones obtained conventionally. Supercritical fluid extraction (SFE) is one of the alternative methods to conventional systems that in the last decade has gained acceptance in the extraction of valuable substances. In order to obtain the best qualitative and quantitative values of bioactive compounds in the extracts, the choosing of the extraction techniques must be followed by the right choice of the solvents and extraction conditions applied. Gwiazdowska et al. 2022 [5] investigated *Glechoma hederacea* var. *longituba* plant extracts obtained by SFE with carbon dioxide (SC-CO2), using methanol as a co-solvent. According to the obtained results, SC-CO2 extracts of *Glechoma* are a promising source of bioactive constituents that can be beneficial in a sustainable manner, acting as natural antioxidants and antibacterial agents. Nevertheless, the biological activity of *G. hederacea* extracts obtained under different conditions of the process was affected by the increasing temperature of the process. The results indicate that the obtained *Glechoma* extracts are characterized by high total phenolic content (TPC) values, which differ depending on the extraction conditions. The high TPC values correlate with high antioxidant properties, as well as antimicrobial (excluding filamentous fungi) and antibiofilm activity (the prevention of biofilm formation) [5].

The research of Giordano et al. 2022 [6] included optimization of the ultrasound-assisted extraction (UAE) of phenolic compounds from kiwi peel, contributing to this fruit waste valorization. This low-cost raw material was determined by its promising bioactive properties, among which antioxidant and antimicrobial effects, and no toxicity to Vero cells were observed. Optimization of the UAE coupled with the response surface methodology (RSM) included different process parameters, including time, ultrasonic power and ethanol concentration. The extraction efficiency was determined gravimetrically and the contents of phenolic compounds identified by HPLC-DAD-ESI/MSn were also used in the optimization. In turn, the polynomial models were fitted to the experimental data and used to determine the optimal conditions. The sonication of the sample under appropriate conditions allowed for the experimental validation of the predictive model. The studies are consistent with the current trends concerning the development of natural ingredients (such as food preservatives) from waste and also the resource-use efficiency and circular bioeconomy [6].

Despite the fact that the Special Issue has been closed, the research concerning the discovery of new sufficient antimicrobial substances, characterized by broad biological properties, is the matter of constant ongoing research. As the newly developed approaches of extraction are investigated, new opportunities for obtaining plant-derived antimicrobials are observed. Furthermore, the application for food products and agriculture of such substances enables the exclusion of their chemical equivalents. Sustainable recovery seems to be an important trend in view of the need to care for the environment. Novel approaches of obtaining antimicrobial substances should be more environmentally friendly and pollution-free; therefore, there is a great need to develop and optimize such methods.

Funding: This research received no external funding.

Acknowledgments: The Special Issue Editors are immensely grateful for the contributions of the hardworking and professional authors. Congratulations to all authors for their interesting works.

Conflicts of Interest: The authors declare no conflict of interest.

References

1. Ziarno, M.; Kozłowska, M.; Ścibisz, I.; Kowalczyk, M.; Pawelec, S.; Stochmal, A.; Szleszyński, B. The Effect of Selected Herbal Extracts on Lactic Acid Bacteria Activity. *Appl. Sci.* **2021**, *11*, 3898. [CrossRef]
2. Gwiazdowska, D.; Marchwińska, K.; Juś, K.; Uwineza, P.A.; Gwiazdowski, R.; Waśkiewicz, A.; Kierzek, R. The Concentration-Dependent Effects of Essential Oils on the Growth of Fusarium graminearum and Mycotoxins Biosynthesis in Wheat and Maize Grain. *Appl. Sci.* **2022**, *12*, 473. [CrossRef]
3. Nguenha, R.; Damyeh, M.S.; Phan, A.D.T.; Hong, H.T.; Chaliha, M.; O'Hare, T.J.; Netzel, M.E.; Sultanbawa, Y. Effect of Photosensitization Mediated by Curcumin on Carotenoid and Aflatoxin Content in Different Maize Varieties. *Appl. Sci.* **2021**, *11*, 5902. [CrossRef]
4. Petropoulos, S.A.; Fernandes, Â.; Calhelha, R.C.; Rouphael, Y.; Petrović, J.; Soković, M.; Ferreira, I.C.F.R.; Barros, L. Antimicrobial Properties, Cytotoxic Effects, and Fatty Acids Composition of Vegetable Oils from Purslane, Linseed, Luffa, and Pumpkin Seeds. *Appl. Sci.* **2021**, *11*, 5738. [CrossRef]

5. Gwiazdowska, D.; Uwineza, P.A.; Frąk, S.; Juś, K.; Marchwińska, K.; Gwiazdowski, R.; Waśkiewicz, A. Antioxidant, Antimicrobial and Antibiofilm Properties of Glechoma hederacea Extracts Obtained by Supercritical Fluid Extraction, Using Different Extraction Conditions. *Appl. Sci.* **2022**, *12*, 3572. [CrossRef]
6. Giordano, M.; Pinela, J.; Dias, M.I.; Calhelha, R.C.; Stojković, D.; Soković, M.; Tavares, D.; Cánepa, A.L.; Ferreira, I.C.F.R.; Caleja, C.; et al. Ultrasound-Assisted Extraction of Flavonoids from Kiwi Peel: Process Optimization and Bioactivity Assessment. *Appl. Sci.* **2021**, *11*, 6416. [CrossRef]

MDPI

Article

The Effect of Selected Herbal Extracts on Lactic Acid Bacteria Activity

Małgorzata Ziarno [1,*], Mariola Kozłowska [2], Iwona Ścibisz [3], Mariusz Kowalczyk [4], Sylwia Pawelec [4], Anna Stochmal [4] and Bartłomiej Szleszyński [5]

1 Division of Milk Technology, Department of Food Technology and Assessment, Institute of Food Science, Warsaw University of Life Sciences–SGGW (WULS–SGGW), 02-787 Warsaw, Poland
2 Department of Chemistry, Institute of Food Science, Warsaw University of Life Sciences–SGGW (WULS–SGGW), 02-787 Warsaw, Poland; mariola_kozlowska@sggw.edu.pl
3 Division of Fruit, Vegetable and Cereal Technology, Department of Food Technology and Assessment, Institute of Food Science, Warsaw University of Life Sciences–SGGW (WULS–SGGW), 02-787 Warsaw, Poland; iwona_scibisz@sggw.edu.pl
4 Department of Biochemistry and Crop Quality, Institute of Soil Science and Plant Cultivation, State Research Institute, 24-100 Puławy, Poland; mkowalczyk@iung.pulawy.pl (M.K.); spawelec@iung.pulawy.pl (S.P.); asf@iung.pulawy.pl (A.S.)
5 Institute of Horticultural Sciences, Warsaw University of Life Sciences–SGGW (WULS–SGGW), 02-787 Warsaw, Poland; bartekszleszynski@gmail.com
* Correspondence: malgorzata_ziarno@sggw.edu.pl; Tel.: +48-225-937-666

Abstract: This study aimed to investigate the effect of plant extracts (valerian *Valeriana officinalis* L., sage *Salvia officinalis* L., chamomile *Matricaria chamomilla* L., cistus *Cistus* L., linden blossom *Tilia* L., ribwort plantain *Plantago lanceolata* L., marshmallow *Althaea* L.) on the activity and growth of lactic acid bacteria (LAB) during the fermentation and passage of milk through a digestive system model. The tested extracts were also characterized in terms of their content of polyphenolic compounds and antioxidant activity. It was observed that the addition of the tested herbal extracts did not inhibit the growth of LAB in fermented milk drinks, such as yogurts. However, they can gradually inhibit fermentation when added at concentrations above 2% by weight, and hence should be used in limited amounts. The microflora of yogurts containing the tested herbal extracts did not die during digestion in model digestive juices, and no stimulating effect of the added plant extracts was noted either. Nevertheless, due to the antioxidant properties, a slight addition of the herbal extracts containing polyphenols to different kinds of food products can increase the nutritional quality, thus making them functional foods.

Keywords: lactic acid bacteria; plant extracts; milk fermentation; total phenolic content; antioxidant capacities

Citation: Ziarno, M.; Kozłowska, M.; Ścibisz, I.; Kowalczyk, M.; Pawelec, S.; Stochmal, A.; Szleszyński, B. The Effect of Selected Herbal Extracts on Lactic Acid Bacteria Activity. *Appl. Sci.* **2021**, *11*, 3898. https://doi.org/10.3390/app11093898

Academic Editor: Gwiazdowska Daniela

Received: 1 April 2021
Accepted: 21 April 2021
Published: 25 April 2021

1. Introduction

The benefits of fermented milk drinks seem to be recognized by consumers as the consumption of these products continues to increase, particularly in industrialized countries. In recent years, manufacturers have been outdoing each other in inventing novel products to attract customers. For instance, yogurts are prepared with various fruits, which are often quite exotic and rarely found in local markets. It has been found that plant extracts that were used in folk medicine and sometimes even in culinary applications may serve as interesting additives. However, the effect of their addition on the microflora of fermented milk beverages should be investigated.

Valerian (*Valeriana officinalis* L.) has been a known herb since ancient Greek and Roman times. The action of this herb was described by Dioscorides, Hippocrates, and Plinius Secundus. Valerian is one of the most popular herbs used in various medications, mainly for its calming and relaxing effects. In folk medicine, it was used as a sedative and as a relaxant of the smooth muscles of the digestive tract, urinary tract, and blood vessels. The name of the herb is derived from the Latin word "valere", which means "to be healthy" [1,2]. Sage (*Salvia officinalis* L.) contains compounds that inhibit lipid peroxidation, and also exhibit antibacterial, antiallergic, antiviral, and analgesic properties. In addition, sage is used in medicine to fight rheumatism and arthritis, and sometimes to prevent the general weakening of the body [3–5]. Chamomile (*Matricaria chamomilla* L.) has been used for centuries as a poultice for healing wounds and burns, and as an eyewash to treat conjunctivitis. This herb has not lost its relevance [6–8]. It is known to have anti-inflammatory, antibacterial, bacterial toxin-binding, disinfectant, antispasmodic, choleretic, sedative, and laxative effects, and is hence used widely in medicine. Furthermore, it is used in the food industry for the production of liqueurs and in the cosmetic industry for the production of creams, lotions, soaps, bath liquids, and toothpastes [6–8]. Cistus (*Cistus* L.) acts as an antioxidant and removes free radicals and prevents the formation of new ones [9,10]; it also has antibacterial, anti-inflammatory, antiviral, and antifungal effects [9,10]. Linden blossom (*Tilia* L.) exhibits a high healing effect and was therefore used by our ancestors for generations. In traditional medicine, the linden flower is commonly used for nervous tension and excessive nervous excitability. In some cases, it is used as a prophylactic against atherosclerosis and hypertension [11–16]. Linden flowers are also characterized by anti-inflammatory, antipyretic, diaphoretic, diuretic, mildly astringent, and analgesic properties [11–16]. Since ancient times, infusions prepared from linden or sage have been used to treat respiratory diseases. Ribwort plantain (*Plantago lanceolata* L.) has also been used in folk medicine. Its leaves are used to treat cuts, bruises, bites, and burns, as well as for chronic gastrointestinal catarrh, acidity, and damage to the gastric and intestinal mucosa (e.g., by bacterial toxins or other chemical compounds) [17–20]. This herb has bacteriostatic, antimicrobial, anti-inflammatory, antispasmodic, and expectorant properties. It also improves blood clotting and has astringent and sealing effects on blood vessels [17–20]. Marshmallow (*Althaea* L.) was valued for its prohealth properties by Egyptians and Syrians in ancient times. The generic name Althaea comes from the Greek word "althe" which means "to heal". Both the root and leaves of marshmallow are rich in mucus, which is the basic ingredient of this plant used in medicine [21–24]. Marshmallow has anti-inflammatory, protective, coating, and antitussive properties [21–24].

In food production, the use of herbs can stabilize the microflora of products, by preventing the development of harmful microorganisms and supporting the growth of desired bacteria. Only few scientific reports have shown that when selected plant extracts are used in appropriate portions, they can have a beneficial effect on lactic acid bacteria (LAB) [25,26]. Bifidobacteria and some lactobacilli can transform polyphenols into important metabolites that have important functions in the human body is known to be [25,26]. Moreover, observations of market trends suggest that plant extracts are used or can be used as food additives in the production of flavored fermented dairy products (including yoghurts or other fermented milks, sour cream, acid and rennet cheeses), and many products derived from them. Only limited studies have analyzed the influence of phenolic compounds on the growth and viability of other lactic acid bacteria such as *Streptococcus thermophilus* used in the production of yogurt.

The present study aimed to investigate the effect of selected herbal additives on the activity and growth of lactic acid bacteria (LAB) during fermentation and passage of milk through the digestive system model. Additionally, the tested extracts were characterized in terms of their content of polyphenolic compounds and antioxidant activity.

2. Materials and Methods

2.1. Materials

The following plant extracts were used in the research: valerian (*V. officinalis* L.), sage (*S. officinalis* L.), chamomile (*M. chamomilla* L.), cistus (*Cistus* L.), linden blossom (*Tilia* L.), ribwort plantain (*P. lanceolata* L.), and marshmallow (*Althaea* L.). They were purchased from GreenVit sp. z.o.o. (Zambrów, Poland). These were water extracts obtained by percolation at elevated temperatures. Then, they were concentrated in a vacuum evaporator, and possibly dried further under vacuum. Marshmallow and plantain extracts were obtained in liquid form, while the rest were in powder form. Maltodextrin was used as the drying carrier.

2.2. Determination of the Effect of Plant Extracts on the Lactic Acid Fermentation of Milk

In the first stage of the work, the ability of LAB to ferment milk was checked in the presence of selected plant extracts. Briefly, the milk samples intended for lactic acid fermentation was prepared from UHT milk (containing 3.2% of fat) in a volume of 100 mL, and the plant extracts were added at amounts of 0.2, 0.6, 1.0, 1.4, 2.0, 3.0, 4.0, and 5.0%. Then, the samples were transferred to a water bath heated to 42 °C, and the starter culture was added (at an amount of 0.04%). The yogurt starter culture YC-X16 (received kindly from Chr. Hansen Poland) was used in the research. This freeze-dried culture is composed of *Streptococcus thermophilus* and *Lactobacillus delbrueckii* subsp. *bulgaricus*. Fermentation of milk samples was carried out at 42 °C. For the next 4 h, the pH of the samples was measured every 30 min until the end of fermentation, using a standard laboratory stationary pH meter with three replications.

2.3. Determination of the Effect of Plant Extracts on the Populations of Lactic Acid Bacteria

The effect of plant extracts on the population of LAB cells was investigated immediately after the fermentation process and digestion of fermented milk in model digestive juices. At this stage, however, based on the results of the first stage, only the following portions of plant extracts were used: 0.2, 0.6, 1.0, and 1.4% extracts added to UHT milk. Fermentation was carried out as in the first stage of the research. After the end of the process, the number of LAB cells in the samples was determined M17 (MERCK) and MRS agar (De Man Rogosa Sharpe Agar, MERCK) were used in the analysis. The inoculated Petri plates were incubated in an incubator at 37 °C under aerobic (M17 agar) or anaerobic (MRS agar) condition. The cell count was determined after 72 h, and the results are expressed in colony forming units in 1 mL of the sample (CFU/mL).

The next stage of the work consisted of several steps. The first step involved the digestion of the fermented milk samples added with plant extracts under gastric juice conditions, and the second step involved the digestion of the samples under intestinal juice conditions [27]. Gastric juice was prepared as described by Ziarno and Zaręba [27]. Briefly, 4.8 g of NaCl, 1.56 g of $NaHCO_3$, 2.2 g of KCl, and 0.22 g of $CaCl_2$ were dissolved in 1000 mL of distilled water. The pH of the prepared solution was adjusted to 2.40 with 1 M HCl. Then, the solution was sterilized in an autoclave at 121 °C for 15 min. Immediately before the experiment, pepsin (Sigma-Aldrich) was added to the solution at an amount of 285 µL/100 mL gastric juice. The model intestinal juice was also prepared as described by Ziarno and Zaręba [27]. Briefly, 5 g of NaCl, 0.6 g of KCl, 0.25 g of $CaCl_2$, and 8.5 g of beef bile were dissolved respectively in 1 M $NaHCO_3$. The pH of the prepared solution was adjusted to 7.0 with 1 M HCl. The whole mixture was successively sterilized in an autoclave at 121 °C for 15 min. Immediately before the experiment, one Kreon Travix 10,000 capsule (Abbott Laboratories) was added to 200 mL of model intestinal juice. The capsule contains a mixture of digestive enzymes, which at a dose of 150 mg shows the activity of 10,000 IU Ph. Eur lipase, 8000 units Ph. Eur amylase, and 600 Ph. Eur proteases. Digestion was carried out in a static system by mixing 35 mL of model gastric juice with the appropriate amount of pepsin and 35 mL of fermented milk sample. Gastric juice digestion was performed for 3 h at 37 °C, and then the mixture was transferred to the same amount of model intestinal juice for digestion which lasted for 5 h at 37 °C. Finally, the number of viable LAB cells was determined as described above.

2.4. Determination of TPC of Plant Extracts

Total phenolic content (TPC) was estimated in the plant extracts using the Folin–Ciocalteu method as described previously [28] with a slight modification. First, appropriately diluted plant extract (3 mg/mL) was mixed with deionized water (20 mL) and Folin–Ciocalteu reagent (0.5 mL). After 30 s, 5 mL of Na_2CO_3 (20%, v/v) was added to the solution. Then, the solution was incubated at 21 °C for 1 h, and its absorbance was measured at 765 nm using a UV/Vis spectrophotometer (Model 8500; Techcomp, Hong Kong). The results were expressed as mg gallic acid equivalents per gram of extract (mg GAE/g extract) using a standard gallic acid calibration curve. The analysis was performed in three independent replications.

2.5. Determination of Antioxidant Capacities of Plant Extracts

Before the analysis of antioxidant activity, each extract (3–6 mg) was diluted with distilled water (2–4 mL). The plant extract solutions were thus prepared in triplicate, and their average values of antioxidant capacity were determined as mmol Trolox equivalents per gram of extract (mmol TE/g extract).

DPPH (2,2-diphenyl-1-picrylhydrazyl) radical scavenging activity was performed according to a procedure described by Yen and Chen [29] with minor modification. Briefly, 1 mL of 0.3 mmol/L freshly prepared DPPH methanol solution was mixed with 0.2 mL of the plant extract solution and 3.8 mL of methanol. The samples were vortex-mixed at high speed for 10 s and incubated for 10 min in the dark at room temperature. Then, their absorbance was measured at 517 nm using a UV/Vis spectrophotometer. A standard curve was obtained using the Trolox standard in the range of 8–40 µmol/L.

ABTS (2,2′-azinobis(3-ethylbenthiazoline-6-sulfonic acid)) radical scavenging activity of the extracts was determined according to the method described by Re et al. [30]. First, the ABTS•+ solution was prepared by mixing ABTS aqueous solution (14 mmol/L) with potassium persulfate aqueous solution (4.9 mmol/L). The prepared solution was kept for 12–16 h in the dark at room temperature. Before the analysis, the ABTS•+ solution was diluted with phosphate-buffered saline (0.01 mol/L, pH 7.4) to achieve an absorbance value of 0.7 ± 0.05 at 734 nm. Then, 40 µL of the plant extract solution or Trolox solution was mixed with 4 mL of ABTS•+ working solution. The reactive mixture was incubated at room temperature in the dark, and after exactly 6 min, its absorbance was recorded at 734 nm. A series of Trolox solutions (final concentrations 0–15 µM) were used for calibration.

FRAP (ferric reducing antioxidant power) assay was performed as described by Benzie and Strain [31] with some modification. Before the assay, the FRAP reagent was freshly prepared by mixing 300 mmol/L acetate buffer (pH 3.6), 20 mmol/L FeCl3 solution, and 10 mmol/L TPTZ (2,4,6-Tris(2-pyridyl)-s-triazine) in 40 mmol/L HCl in a 10:1:1 (v/v/v) proportion and stored away from light. Then, 100 µL of the appropriately diluted sample extract was mixed with 0.3 mL distilled water and 3 mL FRAP reagent. The absorbance of the reaction mixture was measured spectrophotometrically at 593 nm after incubation at 37 °C for 10 min. The blank solution was obtained by mixing 0.3 mL distilled water with 3 mL of FRAP reagent. A standard curve was prepared using Trolox in the range of 80–500 µmol/L. All determinations were carried out in triplicate.

2.6. ESI-QTOF Qualitative Analysis of Plant Extracts

High-resolution liquid chromatography (LC)–mass spectrometry (MS) analyses (exact masses, MS/MS fragmentation patterns, molecular formulas) were performed on a Thermo Scientific Ultimate 3000 RS chromatographic system coupled with a Bruker Impact II HD (Bruker, Billerica, MA, USA) quadrupole time-of-flight (QTOF) mass spectrometer. Chromatographic separations were carried out on a Waters BEH C18 column (2.1 × 150 mm, 1.7 µm; Milford, MA USA), equipped with precolumn. Mobile phase A used was 0.1% (v/v) formic acid, while mobile phase B was acetonitrile containing 0.1% (v/v) of formic acid. A gradient from 7 to 80% of phase B over 30 min was used for separation. The flow rate was set at 0.5 mL/min, and the column was held at a temperature of 60 °C. The injection volume was 5 µL. The light absorption patterns of the investigated sample components were obtained in the wavelength range of 190–600 nm using a photodiode array detector (Thermo Ultimate DAD-3000) with an analytical flow cell. The column's effluent was split into 1:3 proportions between the two detectors operating in parallel, the mass spectrometer, and the charged aerosol detector (CAD), to identify the main constituents of the investigated samples. For identification, data were collected from the mass spectral analyses in both positive and negative ion modes with electrospray ionization (ESI). Linear (centroid) mass spectra were acquired over a mass range from m/z 50 to 2000 with the following MS parameters: positive ion capillary voltage, 4.5 kV; negative ion capillary voltage, 3.0 kV, dry gas (N2) flow, 6 L/min; dry gas temperature, 200 °C; and nebulizer gas (N2) pressure, 0.7 bar. Argon was used as the collision gas. The MS/MS collision energy and parent mass isolation width were automatically set between 2.5 and 35 eV and between 2 and 6 mass units, depending on the m/z of the fragmented ion. The parameters for ion transfer were optimized for m/z 50–2000, with collision cell transfer time at 80 µs and prepulse storage at 10 µs. The acquired data were calibrated internally with 10 mM sodium formate introduced to the ion source via a 20-µL loop at the beginning of each separation process. Data processing was carried out using Bruker DataAnalysis 4.3 software. The main components of the sample were identified from the CAD peak areas, while the constituents were identified based on the light absorption properties, precise mass measurements (measurement error <5 ppm) of the primary ion m/z, which allowed for calculating the molecular formula, and software-aided analysis of the isotopic and MS/MS fragmentation patterns [32].

2.7. Statistical Analysis

Multifactor analysis of variance (ANOVA) is used in statistical analysis to determine the influence of significant factors in a multivariate model. This is a typical system used for experimental analysis which, in addition to checking the influence of one factor, allows checking the interaction of individual factors with each other. In this study, in conjunction with ANOVA, Tukey's test was used at a significance level of 0.05 to analyze the mutual influence of two factors on each other and find those that differ significantly from each other, which shows the pairs that are statistically significant.

3. Results

3.1. Determination of the Effect of Plant Extracts on Lactic Acid Fermentation

In the first stage of the research, changes in the pH of milk samples enriched with portions of plant extracts were analyzed. The obtained results are presented in Figure 1. The change in pH during the fermentation process indicated that none of the added plant extracts inhibited fermentation. Statistical analysis showed that only the milk samples added with the first three doses of the extracts (from 0 to 1.0%) constituted one homogeneous group ($p = 0.001$), where the pH values were higher by on average 0.2–0.3 than the samples added with higher doses of the tested plant extracts. The milk samples with 1.0% or higher portion of the plant extracts constituted separate homogeneous groups ($p = 0.001$) for each analyzed extract. It should be noted that the pH did not reduce below 4.5 for the addition of any extract above 2.0% concentration. For selected additives (sage and cistus), a pH of even 4.1 was achieved for the lowest dose, and 4.8 for 5.0% dose. For the remaining additives, the differences in pH observed between their lowest and the highest portion were approximately 0.4. This difference is significant and confirms that a large amount of additives can inhibit the biochemical activity (i.e., acidifying activity) of LAB present in the yogurt starter culture. In all cases of milk samples fermented in the presence of the tested plant extracts, the pH was no longer statistically significantly reduced after 3.5 h of fermentation.

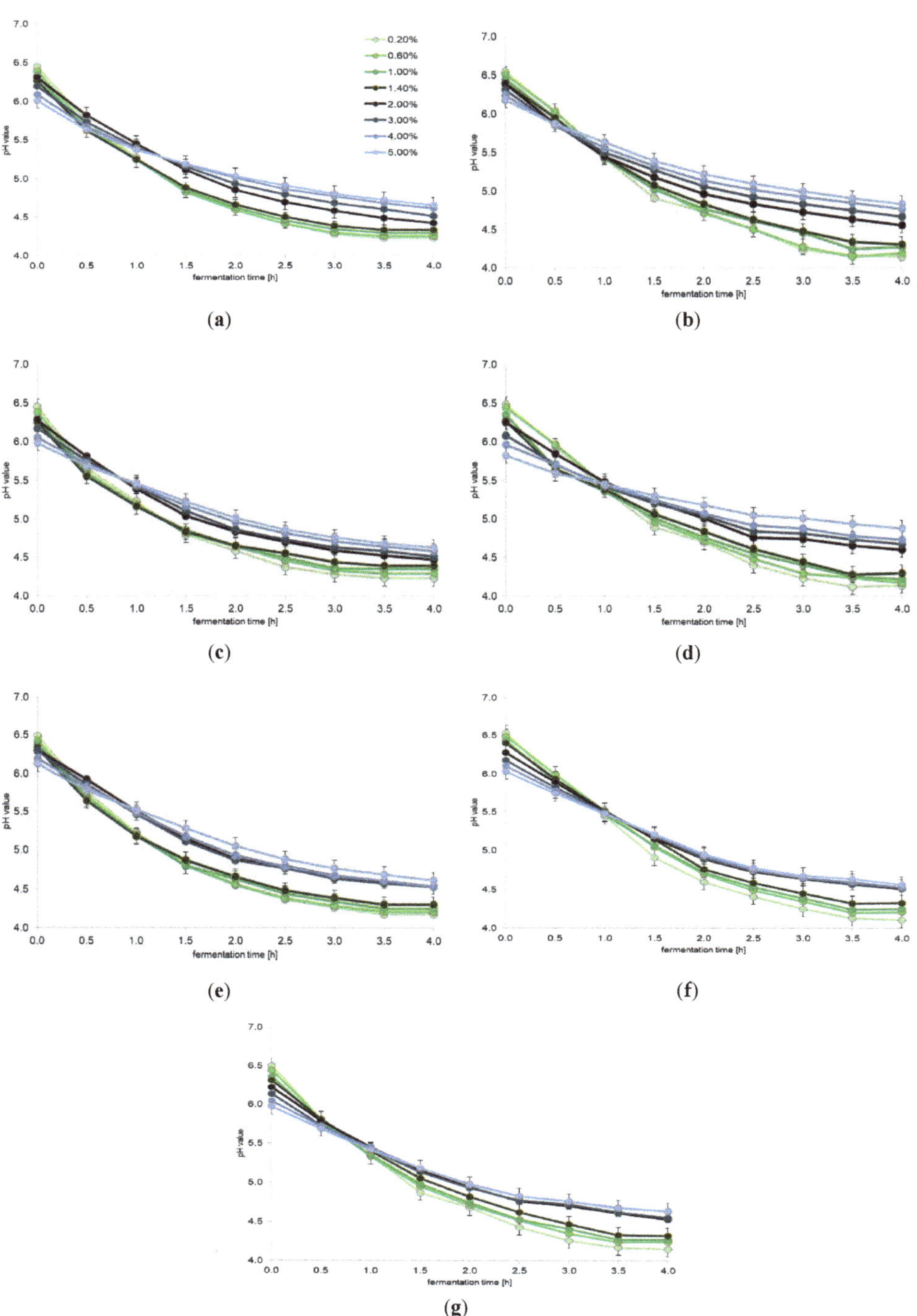

Figure 1. The effect of plant extracts on lactic acid fermentation (average values and standard deviations of three replicates): (**a**) valerian (*Valeriana officinalis* L.); (**b**) sage (*Salvia officinalis* L.); (**c**) chamomile (*Matricaria chamomilla* L.); (**d**) cistus (*Cistus* L.); (**e**) linden blossom (*Tilia* L.); (**f**) ribwort plantain (*Plantago lanceolata* L.); and (**g**) marshmallow (*Althaea* L.).

3.2. Determination of the Effect of Plant Extracts on the Populations of Lactic Acid Bacteria

The results of the determination of both tested LAB populations (Table 1) immediately after the end of the fermentation process confirmed that most of the tested plant extracts did not reduce the populations ($p < 0.05$). This means that the tested LAB cultures well tolerated the presence of the extracts at an amount of up to 3.0% in milk. Only with the addition of sage extract, we observed some slight, but statistically significant, reduction in the *Lactobacillus* cell population (Table 1).

Further analyses showed that the types of additives and the digestion process had an impact on the population of *S. thermophilus* cells ($p < 0.05$); however, no effect on the lactobacilli population was observed ($p > 0.05$). It should be noted that both bacterial species had different survival rates in the samples and reacted differently to digestion in the model digestive juices (Table 1). Statistical analysis carried out for individual bacterial species showed that only digestion influenced the number of *S. thermophilus* cells in most cases. The exceptions were the milk samples fermented with the addition of cistus extract and marshmallow extract, in which the digestion step did not statistically significantly influence the *S. thermophilus* cell population. This may suggest that these two extracts improved the survival of *S. thermophilus* cells under the conditions of the model digestive system. In the case of *Lactobacillus* bacteria, neither the type of plant extract or its dose nor digestion in the model digestive system had a statistically significant influence on the bacterial cell population (Table 1).

Table 1. The effect of plant extracts on the populations of lactic acid bacteria (average values and standard deviations of five replicates).

Plant Extracts	Sampling Time	Additive Level						Additive Level					
		0.2%	0.6%	1.0%	1.4%	2.0%	3.0%	0.2%	0.6%	1.0%	1.4%	2.0%	3.0%
		Streptococcus thermophilus Population [log CFU/mL]						*Lactobacillus delbrueckii* subsp. *bulgaricus* Population [log CFU/mL]					
Valerian (*Valeriana officinalis* L.)	after fermentation	7.9 ± 0.3 ^{a}	7.6 ± 0.5 ^{a}	7.8 ± 0.3 ^{a}	7.8 ± 0.3 ^{a}	7.6 ± 0.3 ^{a}	7.4 ± 0.3 ^{a}	7.2 ± 0.3 ^{a}	7.1 ± 0.5 ^{a}	7.0 ± 0.5 ^{a}	7.0 ± 0.3 ^{a}	6.6 ± 0.3 ^{a}	6.4 ± 0.3 ^{a}
	after digestion	6.6 ± 0.3 ^{b}	6.5 ± 0.3 ^{b}	7.0 ± 0.3 ^{b}	6.8 ± 0.3 ^{b}	6.5 ± 0.3 ^{b}	6.3 ± 0.3 ^{b}	6.7 ± 0.3 ^{a}	6.5 ± 0.3 ^{a}	6.8 ± 0.3 ^{a}	6.9 ± 0.4 ^{a}	6.5 ± 0.3 ^{a}	6.1 ± 0.3 ^{a}
Sage (*Salvia officinalis* L.)	after fermentation	7.8 ± 0.3 ^{a}	8.0 ± 0.3 ^{a}	7.7 ± 0.4 ^{a}	7.9 ± 0.2 ^{a}	7.7 ± 0.3 ^{a}	7.3 ± 0.4 ^{a,b}	7.2 ± 0.3 ^{a}	6.7 ± 0.3 ^{a,b}	6.4 ± 0.3 ^{b}	6.4 ± 0.3 ^{b}	6.4 ± 0.3 ^{b}	6.5 ± 0.3 ^{a,b}
	after digestion	6.9 ± 0.3 ^{b}	7.0 ± 0.2 ^{b}	6.3 ± 0.3 ^{b,c}	6.8 ± 0.3 ^{b,c}	6.6 ± 0.3 ^{b,c}	6.2 ± 0.3 ^{c}	6.8 ± 0.3 ^{a,b}	6.3 ± 0.4 ^{b}	6.2 ± 0.4 ^{b}	6.2 ± 0.4 ^{b}	6.1 ± 0.3 ^{b}	6.1 ± 0.3 ^{b}
Chamomile (*Matricaria chamomilla* L.)	after fermentation	7.9 ± 0.3 ^{a}	7.9 ± 0.3 ^{a}	8.0 ± 0.3 ^{a}	7.9 ± 0.3 ^{a}	7.7 ± 0.3 ^{a}	7.5 ± 0.3 ^{a}	7.2 ± 0.4 ^{a}	7.0 ± 0.5 ^{a}	6.8 ± 0.5 ^{a}	6.8 ± 0.3 ^{a}	6.5 ± 0.3 ^{a}	6.4 ± 0.3 ^{a}
	after digestion	6.8 ± 0.3 ^{b}	6.3 ± 0.3 ^{b}	6.9 ± 0.3 ^{b}	6.9 ± 0.3 ^{b}	6.5 ± 0.3 ^{b}	6.3 ± 0.3 ^{b}	6.6 ± 0.4 ^{a}	6.3 ± 0.4 ^{a}	6.6 ± 0.4 ^{a}	6.7 ± 0.4 ^{a}	6.5 ± 0.5 ^{a}	6.2 ± 0.4 ^{a}
Cistus (*Cistus* L.)	after fermentation	8.1 ± 0.3 ^{a}	8.0 ± 0.3 ^{a}	7.9 ± 0.3 ^{a}	8.0 ± 0.3 ^{a}	7.8 ± 0.3 ^{a,b}	7.4 ± 0.4 ^{a,b}	7.2 ± 0.5 ^{a}	7.2 ± 0.3 ^{a}	7.2 ± 0.3 ^{a}	7.6 ± 0.3 ^{a}	7.2 ± 0.3 ^{a}	6.8 ± 0.3 ^{a}
	after digestion	7.9 ± 0.4 ^{a}	7.8 ± 0.4 ^{a,b}	7.8 ± 0.4 ^{a,b}	7.5 ± 0.3 ^{a,b}	7.6 ± 0.3 ^{a,b}	7.1 ± 0.2 ^{b}	6.7 ± 0.4 ^{a}	7.0 ± 0.4 ^{a}	7.0 ± 0.4 ^{a}	7.5 ± 0.4 ^{a}	7.0 ± 0.4 ^{a}	6.6 ± 0.4 ^{a}
Linden blossom (*Tilia* L.)	after fermentation	8.2 ± 0.3 ^{a}	8.3 ± 0.3 ^{a}	8.3 ± 0.3 ^{a}	8.2 ± 0.3 ^{a}	7.9 ± 0.5 ^{a}	7.8 ± 0.5 ^{a}	7.5 ± 0.4 ^{a}	7.7 ± 0.3 ^{a}	7.2 ± 0.3 ^{a}	7.5 ± 0.3 ^{a}	7.3 ± 0.3 ^{a}	6.9 ± 0.3 ^{a,b}
	after digestion	6.9 ± 0.3 ^{b}	6.8 ± 0.3 ^{b}	6.9 ± 0.3 ^{b}	6.8 ± 0.3 ^{b}	6.6 ± 0.3 ^{b}	6.4 ± 0.3 ^{b}	7.0 ± 0.3 ^{a}	6.8 ± 0.3 ^{a,b}	7.0 ± 0.3 ^{a}	6.7 ± 0.3 ^{a,b}	6.7 ± 0.3 ^{a,b}	6.3 ± 0.3 ^{b}
Ribwort plantain (*Plantago lanceolata* L.)	after fermentation	8.1 ± 0.3 ^{a}	8.1 ± 0.4 ^{a}	8.3 ± 0.3 ^{a}	8.3 ± 0.3 ^{a}	7.9 ± 0.2 ^{a}	7.5 ± 0.5 ^{a,b}	7.3 ± 0.5 ^{a}	7.1 ± 0.3 ^{a}	7.1 ± 0.3 ^{a}	7.3 ± 0.3 ^{a}	7.2 ± 0.3 ^{a}	6.8 ± 0.3 ^{a}
	after digestion	7.1 ± 0.3 ^{b}	6.7 ± 0.3 ^{b,c}	6.6 ± 0.3 ^{b,c}	6.5 ± 0.3 ^{b,c}	6.5 ± 0.3 ^{b,c}	6.2 ± 0.3 ^{c}	7.0 ± 0.4 ^{a}	6.7 ± 0.3 ^{a}	7.0 ± 0.4 ^{a}	7.2 ± 0.3 ^{a}	6.8 ± 0.3 ^{a}	6.5 ± 0.3 ^{a}
Marshmallow (*Althaea* L.)	after fermentation	8.2 ± 0.3 ^{a}	8.1 ± 0.3 ^{a}	8.0 ± 0.3 ^{a}	8.0 ± 0.3 ^{a}	7.8 ± 0.3 ^{a}	7.4 ± 0.2 ^{a}	7.7 ± 0.5 ^{a}	7.4 ± 0.3 ^{a}	7.5 ± 0.4 ^{a}	7.8 ± 0.3 ^{a}	7.4 ± 0.3 ^{a}	7.0 ± 0.3 ^{a,b}
	after digestion	7.2 ± 0.3 ^{a,b}	7.2 ± 0.3 ^{a,b}	7.2 ± 0.3 ^{a,b}	7.3 ± 0.3 ^{a}	7.1 ± 0.3 ^{a,b}	6.8 ± 0.3 ^{b}	7.2 ± 0.3 ^{a}	7.1 ± 0.3 ^{a,b}	7.1 ± 0.3 ^{a,b}	7.3 ± 0.3 ^{a}	7.0 ± 0.3 ^{a,b}	6.6 ± 0.3 ^{b}

^{a,b,c} the same letter indices within a given bacterial species and for a given plant extract mean no statistically significant differences at the significance level of 0.05.

3.3. TPC and Antioxidant Capacities of Plant Extracts

The phenolic content determined in the tested plant extracts is presented in Table 2. A significant difference was observed in the phenolic content between cistus and linden blossom extracts and the rest of the tested plant extracts. As we mentioned in the materials and methods section marshmallow and plantain extracts were in liquid form while the rest were in powder form. As can be seen from the data presented in Table 2, liquid marshmallow and plantain extracts were not preparations with the lowest total phenolic content, as well as antioxidant capacity. Statistical analysis showed significant differences in the phenolic content, which resulted in the identification of seven homogenous groups (at a 5% significance level).

Table 2. The TPC and Trolox equivalent antioxidant capacity of plant extracts determined by DPPH, ABTS, and FRAP assays (average values and standard deviations of three replicates).

Plant Extracts	Total Phenolic Content [mg GAE/g of Extract]	Antioxidant Capacities [mmol TE/g Extract]		
		DPPH	ABTS	FRAP
Valerian (*Valeriana officinalis* L.)	30.97 ± 0.49 [f]	0.104 ± 0.003 [b]	0.159 ± 0.003 [b]	0.090 ± 0.004 [a]
Sage (*Salvia officinalis* L.)	61.42 ± 0.43 [c]	0.137 ± 0.002 [c]	0.186 ± 0.004 [c]	0.122 ± 0.004 [b,c]
Chamomile (*Matricaria chamomilla* L.)	32.56 ± 0.24 [e]	0.099 ± 0.003 [b]	0.120 ± 0.003 [a]	0.092 ± 0.004 [a]
Cistus (*Cistus* L.)	106.38 ± 0.01 [a]	0.154 ± 0.005 [d]	0.208 ± 0.004 [d]	0.136 ± 0.003 [d]
Linden blossom (*Tilia* L.)	104.72 ± 0.39 [b]	0.161 ± 0.007 [d]	0.210 ± 0.007 [d]	0.133 ± 0.004 [c,d]
Ribwort plantain (*Plantago lanceolata* L.)	41.84 ± 0.20 [d]	0.129 ± 0.004 [c]	0.150 ± 0.003 [b]	0.113 ± 0.006 [b]
Marshmallow (*Althaea* L.)	24.06 ± 0.34 [g]	0.084 ± 0.003 [a]	0.123 ± 0.005 [a]	0.098 ± 0.007 [a]

Different letters ([a–g]) within the same column indicate significant difference at the significance level of 0.05.

The plant extracts were also screened for DPPH, ABTS, and FRAP radical scavenging activities. It was observed that cistus and linden blossom extracts exhibited the highest DPPH, ABTS, and FRAP radical scavenging activity, whereas the lowest DPPH, ABTS, and FRAP activity was exhibited by marshmallow, chamomile, and valerian extract, respectively. Correlation analysis between TPC and DPPH radical scavenging ability, TPC and ABTS radical scavenging ability, and TPC and FRAP radical scavenging ability showed a high degree of correlation (r2 = 0.930, 0.923, and 0.931, respectively). However, the r2 values of 0.931, 0.936, and 0.883 determined for correlation between DPPH and ABTS radical scavenging activity, DPPH and FRAP radical scavenging activity, and ABTS and FRAP radical scavenging activity, respectively, indicated that mainly phenolic compounds contributed to the total antioxidant activity in the tested plant extracts.

3.4. Qualitative Analysis of Plant Extracts

The results of ultrahigh-performance LC (UHPLC)-QTOF-CAD analyses of plant extracts are shown in Tables 3–9. The identified compounds are presented according to their elution order. Compounds were tentatively identified based on their HRMS and HRMS2 spectra. The fragmentation patterns and molecular formulas of the compounds were compared with available literature data.

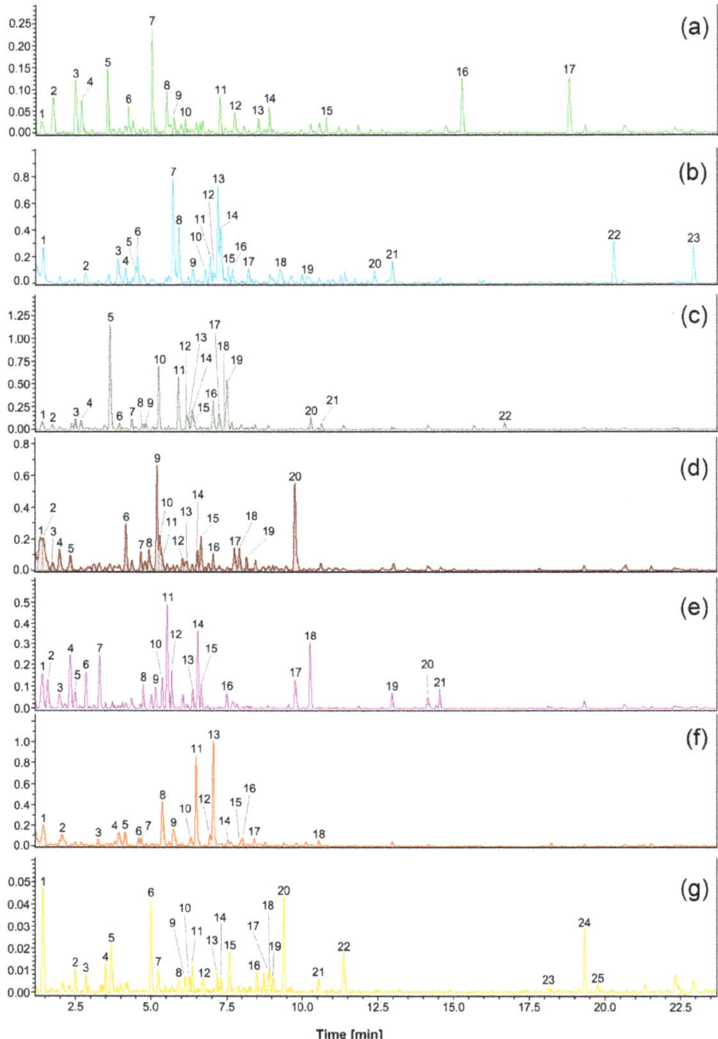

Figure 2. UHPLC-QTOF-CAD chromatograms of the studied plant extracts: (**a**) valerian (*Valeriana officinalis* L.); (**b**) sage (*Salvia officinalis* L.); (**c**) chamomile (*Matricaria chamomilla* L.); (**d**) cistus (*Cistus* L.); (**e**) linden blossom (*Tilia* L.); (**f**) ribwort plantain (*Plantago lanceolata* L.); and (**g**) marshmallow (*Althaea* L.).

Table 3. Compounds identified in the valerian (*Valeriana officinalis* L.) extract using UHPLC-QTOF-MS/MS. Compound numbers correspond to those indicated in Figure 2.

Peak	RT [min]	Molecular Ion [M-H]⁻	Ion Mode	MS/MS Fragments	Formula	Error [ppm]	mSigma	Tentative Identification
1	1.4	255.0510 449.1298	- -	255(100), 193(10) 221(100)	$C_{11}H_{11}O_7$ $C_{18}H_{25}O_{13}$	0.2 0.6	13.2 43.8	piscidic acid unidentified
2	1.8	361.1504	-	361(100), 199(5)	$C_{16}H_{25}O_9$	−0.1	11.3	unidentified
3	2.5	351.0717	-	191(100), 215(11)	$C_{16}H_{15}O_9$	1.3	20.1	caffeoylquinic acid (quinone form)
4	2.7	353.0873	-	173(100), 191(64), 179(45)	$C_{16}H_{17}O_9$	1.4	11.0	neochlorogenic acid
5	3.5	553.1929	-	391(100), 195(21)	$C_{26}H_{33}O_{13}$	−0.4	27.4	unidentified hexoside (lignan)
6	4.2	697.2346 553.1916	- -	373(100), 535(14), 181(11) 391(100), 195(62)	$C_{32}H_{41}O_{17}$ $C_{26}H_{33}O_{13}$	0.4 1.9	15.4 118.8	hydroxypinoresinol di-hexoside unidentified hexoside (lignan)
7	5.0	681.2390	-	357(100)	$C_{32}H_{41}O_{16}$	1.5	2.6	pinoresinol di-hexoside
8	5.5	637.2348	-	197(100), 221(52), 341(31)	$C_{27}H_{41}O_{17}$	0.2	10.4	kanokoside C isomer
9	5.7	535.1814	-	373(100), 181(45)	$C_{26}H_{31}O_{12}$	1.4	4.8	hydroxypinoresinol hexoside
10	6.1	493.2285	-	493(100), 331(47)	$C_{22}H_{37}O_{12}$	1.1	16.2	rhodioloside isomer
11	7.3	519.1869	-	357(100), 151(8)	$C_{26}H_{31}O_{11}$	0.6	7.1	pinoresinol hexoside
12	7.8	609.1822 457.1709	- -	301(100) 293(100)	$C_{28}H_{33}O_{15}$ $C_{21}H_{29}O_{11}$	0.5 1.4	22.0 76.6	hesperidin isomer unidentified
13	8.5	345.1552 347.1708	- -	345(100) 347(100)	$C_{16}H_{25}O_8$ $C_{16}H_{27}O_8$	0.7 1.1	4.4 30.1	uidentified monoterpene hexoside (iridoid) uidentified monoterpene hexoside (iridoid)
14	8.9	477.2339	-	477(100), 315(35)	$C_{22}H_{37}O_{11}$	0.5	6.4	unidentified monoterpene diglycoside
15	10.8	573.2553	-	573(100), 231(4)	$C_{27}H_{41}O_{13}$	−0.1	2.3	unidentified
16	15.3	249.1488	-	249(100), 163(4)	$C_{15}H_{21}O_3$	3.4	0.7	valerenolic acid
17	18.8	291.1592	-	291(100), 249(19)	$C_{17}H_{23}O_4$	3.4	3.1	acetylvalerenolic acid

Table 4. Compounds identified in the sage (*Salvia officinalis* L.) extract using UHPLC-QTOF-MS/MS. Compound numbers correspond to those indicated in Figure 2.

Peak	RT [min]	Molecular Ion [M-H]⁻	Ion Mode	MS/MS Fragments	Formula	Error [ppm]	mSigma	Tentative Identification
1	1.4	197.0458	-	179(53), 135(24), 123(23), 195(6), 151(4)	$C_9H_9O_5$	−1.3	10.2	danshensu isomer
2	2.8	325.0933	-	179(100), 135(14)	$C_{15}H_{17}O_8$	−1.9	11.8	caffeic acid-deoxyhexoside
3	3.9	389.1789	+	227(100), 209(98), 191(56), 131(12)	$C_{18}H_{29}O_9$	4.3	13.8	unidentified
4	4.1	355.1041 583.2047	- -	295(100), 265(51), 160(17), 193(16), 175(10) 373(100), 361(98), 298(58), 295(33), 313(27)	$C_{16}H_{19}O_9$ $C_{27}H_{35}O_{14}$	−1.9 −2.5	19.8 12.4	ferulic acid-hexoside unidentified
5	4.5	493.0628	-	299(100), 271(50), 241(9)	$C_{21}H_{17}O_{14}$	−0.9	12.2	unidentified

Table 4. *Cont.*

Peak	RT [min]	Molecular Ion [M-H]⁻	Ion Mode	MS/MS Fragments	Formula	Error [ppm]	mSigma	Tentative Identification
6	4.6	227.1275	+	209(100), 191(87), 149(55), 167(43), 131(41)	$C_{12}H_{19}O_4$	1.4	0.6	unidentified
7	5.7	461.0731	-	285(100), 255(33)	$C_{21}H_{17}O_{12}$	−1.2	1.5	luteoline-oxyhexoside
8	5.9	593.1522 447.0938	- -	285(100), 339(12), 255(5) 284(100), 256(7)	$C_{27}H_{29}O_{15}$ $C_{21}H_{19}O_{11}$	−1.7 −1.1	6.5 2.4	luteoline-hexoside-deoxyhexoside luteoline-hexoside
9	6.4	551.1770	-	235(100), 533(88), 295(69), 160(64)	$C_{26}H_{31}O_{13}$	0.1	20.5	unidentified
10	6.8	577.1200	-	269(100)	$C_{26}H_{25}O_{15}$	−0.2	11.6	apigenin-hexoside-deoxyhexoside
11	7.0	445.0779	-	269(100)	$C_{21}H_{17}O_{11}$	−0.6	9.0	apigenin-oxyhexoside
12	7.1	431.0984	-	268(100)	$C_{21}H_{19}O_{10}$	0.0	3.3	apigenin-hexoside
13	7.2	359.0768	-	161(100), 197(77), 179(32)	$C_{18}H15O_8$	1.1	4.3	rosmarinic acid
14	7.3	555.1141	-	359(100), 161(22), 135(16), 193(14)	$C_{27}H_{23}O_{13}$	0.7	7.9	salvianolic acid K isomer
15	7.5	475.0881	-	284(100), 299(61)	$C_{22}H_{19}O_{12}$	0.2	24.7	hispiludin/diosmetin-oxyhexoside
16	7.7	463.1224 609.1771	+ +	301(100) 301(100), 463(4)	$C_{22}H_{23}O_{11}$ $C_{35}H_{29}O_{10}$	2.3 −2.7	8.8 44.0	kaempferide-hexoside kaempferide-hexoside-deoxyhexoside
17	8.2	769.1637	-	285(100), 255(26), 575(4)	$C_{36}H_{33}O_{19}$	−2.1	22.5	luteolin-oxyhexoside-pentoside-ferulic acid
18	9.3	753.1679 621.1842	-	269(100), 486(7), 193(4) 313(100), 297(44)	$C_{36}H_{33}O_{18}$ $C_{29}H_{33}O_{15}$	−0.9 −2.7	18.8 14.2	apigenin-oxyhexoside-pentoside-ferulic acid unidentified
19	10.2	711.3968	-	503(100), 453(10)	$C_{37}H_{59}O_{13}$	−1.0	19.4	unidentified
20	12.4	493.1140	-	359(100), 323(40), 135(21), 179(16), 295(14)	$C_{26}H_{21}O_{10}$	0.0	13.0	unidentified
21	13.0	327.2178	-	327(100), 211(12), 229(5), 171(3)	$C_{18}H_{31}O_5$	−0.3	6.2	unidentified
22	20.3	329.1760	-	285(100)	$C_{20}H_{25}O_4$	−0.4	13.3	carnosol isomer
23	22.9	331.1926	-	287(100)	$C_{20}H_{27}O_4$	−3.3	11.9	carnosoic acid isomer

Table 5. Compounds identified in the chamomile (*Matricaria chamomilla* L.) extract using UHPLC-QTOF-MS/MS. Compound numbers correspond to those indicated in Figure 2.

Peak	RT [min]	Molecular Ion [M-H]⁻	Ion Mode	MS/MS Fragments	Formula	Error [ppm]	mSigma	Tentative Identification
1	1.4	315.0719 329.0874	- -	315(100), 152(10) 167(100), 329(28)	$C_{13}H_{15}O_9$ $C_{14}H_{17}O_9$	0.7 1.2	3.8 18.3	protocatechuoyl-hexoside vanilic acid-hexoside
2	1.7	353.0877	-	191(100), 179(40), 135(30)	$C_{16}H_{17}O_9$	0.2	1.6	chlorogenic acid
3	2.5	353.0875	-	191(100)	$C_{16}H_{17}O_9$	1.0	7.7	caffeoylqunic acid
4	2.7	353.0874	-	191(100), 173(83)	$C_{16}H_{17}O_9$	1.3	4.5	caffeoylqunic acid
5	3.7	355.1035 515.1201	- -	193(100), 149(44) 353(100), 191(59), 179(22), 135(12)	$C_{16}H_{19}O_9$ $C_{25}H_{23}O_{12}$	0.0 −1.2	5.5 34.5	ferulic acid hexoside dicaffeoylqunic acid

Table 5. *Cont.*

Peak	RT [min]	Molecular Ion [M-H]$^-$	Ion Mode	MS/MS Fragments	Formula	Error [ppm]	mSigma	Tentative Identification
6	4.0	639.1564	-	313(100), 477(53), 270(36)	$C_{28}H_{31}O_{17}$	0.4	11.4	isorhamnetin-di-hexoside
7	4.4	479.0834 609.1464	- -	317(100), 287(10), 165(6) 284(100), 447(38)	$C_{21}H_{19}O_{13}$ $C_{27}H_{29}O_{16}$	−0.7 −0.4	15.1 28.5	myricetin-oxyhexoside luteolin-di-hexoside
8	4.7	367.1037 463.0883	- -	367(100), 173(19), 193(9), 134(7) 300(100), 227(8)	$C_{17}H_{19}O_9$ $C_{21}H_{19}O_{12}$	−0.6 −0.2	6.5 9.7	feruloylquinic acid quercetin- hexoside
9	4.8	609.1469	-	301(100)	$C_{27}H_{29}O_{16}$	−1.3	5.3	quercetin-hexoside-deoxyhexoside
10	5.3	355.1039	-	193(100), 149(27)	$C_{16}H_{19}O_9$	−1.3	9.5	ferulic acid hexoside
11	5.9	593.1516 447.0937	- -	285(100) 284(100)	$C_{27}H_{29}O_{15}$ $C_{21}H_{19}O_{11}$	−0.6 −1.0	5.0 3.7	kaempferol-hexoside-deoxyhexoside kaempferol-hexoside
12	6.2	493.1002	-	331(100), 287(83), 315(55)	$C_{22}H_{21}O_{13}$	−2.9	6.4	petuletin-hexoside
13	6.2	467.1679	-	323(100), 305(66), 189(25)	$C_{26}H_{27}O_8$	6.8	28.6	unidentified
14	6.4	623.1624 515.1201	- -	315(100), 300(88), 271(28) 353(100), 191(36), 179(26), 135(11)	$C_{28}H_{31}O_{16}$ $C_{25}H_{23}O_{12}$	−1.1 −1.3	29.1 5.4	isorhamnetin-hexoside-deoxyhexoside dicaffeoylqunic acid
15	6.5	681.1674 515.1202	-	313(100), 270(45), 519(27), 477(24) 353(100), 191(71), 179(25)	$C_{30}H_{33}O_{18}$ $C_{25}H_{23}O_{12}$	−0.3 −1.3	13.3 25.4	isorhamnetin-hexoside-acylhexoside dicaffeoylqunic acid
16	7.1	431.0991	-	268(100)	$C_{21}H_{19}O_{10}$	−1.6	7.7	apigenin-hexoside
17	7.3	515.1204 445.1727	- -	353(100), 179(18), 191(15), 135(6) 445(100), 243(44), 183(6), 139(3)	$C_{25}H_{23}O_{12}$ $C_{20}H_{29}O_{11}$	−1.7 −2.5	7.2 22.4	dicaffeoylqunic acid unidetified
18	7.5	479.1169 609.1773	+ +	317(100) 301(100), 463(4)	$C_{29}H_{19}O_7$ $C_{35}H_{29}O_{10}$	−9.1 −2.9	14.9 41.0	isorhamnetin-hexoside kaempferide-hexoside-deoxyhexoside
19	7.5	477.1045 711.2156	- -	477(100), 299(65), 271(63), 315(46) 711(100), 549(20), 433(10), 271(66)	$C_{22}H_{21}O_{12}$ $C_{32}H_{39}O_{18}$	−1.4 −2.0	10.3 9.1	isorhamnetin-hexoside unidentified-hexoside deoxyhexoside-pentoside (flavonoid)
20	10.3	475.1216	+	271(100)	$C_{23}H_{23}O_{11}$	3.9	7.4	apigenin-acylhexoside
21	10.6	519.1141	-	271(72), 299(46), 313(16), 151(7)	$C_{24}H_{23}O_{13}$	0.5	1.9	isorhamnetin-acylhexoside
22	16.7	447.2009	+	219(100), 181(65), 231(30), 411(25), 358(23)	$C_{24}H_{31}O_8$	1.0	8.8	unidentified

Table 6. Compounds identified in the cistus (*Cistus* L.) extract using UHPLC-QTOF-MS/MS. Compound numbers correspond to those indicated in Figure 2.

Peak	RT [min]	Molecular Ion [M-H]⁻	Ion Mode	MS/MS Fragments	Formula	Error [ppm]	mSigma	Tentative Identification
1	1.4	305.0667	-	305(100), 219(12), 165(8)	$C_{15}H_{13}O_7$	−0.1	5.1	gallocatechin isomer
2	1.4	469.0054	-	425(100), 299(24)	$C_{21}H_9O_{13}$	−1.0	12.5	valoneic acid dilacton isomer
3	1.8	633.0743	-	301(100), 275(49), 229(23), 257(22)	$C_{27}H_{21}O_{18}$	−1.5	1.1	strictinin isomer
4	2.0	305.0672 591.1370	- -	305(100), 219(13), 261(8), 179(8) 285(100), 305(12)	$C_{15}H_{13}O_7$ $C_{27}H_{27}O_{15}$	−1.9 −2.5	5.4 41.7	gallocatechin isomer unidentified
5	2.3	289.0722 633.0742	- -	289(100), 245(17), 203(9) 301(100), 275(41), 257(22), 229(11)	$C_{15}H_{13}O_6$ $C_{27}H_{21}O_{18}$	−1.6 −1.4	1.1 35.6	epicatechin strictinin isomer
6	4.2	479.0846	-	316(100), 271(83)	$C_{21}H_{19}O_{13}$	−3.2	13.7	myricetin-hexoside
7	4.7	327.1458	-	327(100), 165(10)	$C_{16}H_{23}O_7$	−2.8	9.2	unidentified
8	5.0	449.0738	-	316(100), 271(68)	$C_{20}H_{17}O_{12}$	−2.8	4.1	myricetin-pentoside
9	5.2	463.0897	-	316(100), 271(91), 179(2)	$C_{21}H_{19}O_{12}$	−3.2	3.5	myricetin-deoxyhexoside
10	5.3	463.0898	-	271(100), 300(83)	$C_{21}H_{19}O_{12}$	−3.6	0.8	quercetin-hexoside
11	5.4	609.1474	-	271(100), 300(76)	$C_{27}H_{29}O_{16}$	−2.1	6.5	quercetin-hexoside-deoxyhexoside
12	6.1	433.0785	-	300(100), 271(86), 255(35), 243(18)	$C_{20}H_{17}O_{11}$	−2.0	5.8	quercetin-pentoside
13	6.2	449.1075	+	197(100), 287(34), 179(11)	$C_{21}H_{21}O_{11}$	0.9	11.9	unidentified
14	6.5	577.1573 447.0944	- -	283(100), 255(57), 285(55), 431(24) 255(100), 227(87), 284(49)	$C_{27}H_{29}O_{14}$ $C_{21}H_{19}O_{11}$	−1.8 −2.6	15.6 11.4	kaempferol-di-deoxyhexoside kaempferol-hexoside
15	6.7	773.1905	+	147(100), 319(18)	$C_{36}H_{37}O_{19}$	2.4	44.2	myricetin-di-deoxyhexoside-hexoside
16	7.1	507.2219	+	219(100), 189(5)	$C_{26}H_{35}O_{10}$	1.1	9.2	unidentified
17	7.8	523.2199	-	475(100), 327(17), 149(7)	$C_{26}H_{35}O_{11}$	−2.7	3.3	unidentified
18	7.9	627.1315	+	147(100), 319(58)	$C_{30}H_{27}O_{15}$	4.8	37.4	myricetin-hexoside-deoxyhexoside
19	8.2	551.2145	-	329(100), 269(36), 314(35)	$C_{27}H_{35}O_{12}$	−1.9	7.2	unidentified
20	9.8	595.1432	+	147(100), 287(39)	$C_{30}H_{27}O_{13}$	2.5	13.1	unidentified

Table 7. Compounds identified in the linden blossom (*Tilia* L.) extract using UHPLC-QTOF-MS/MS. Compound numbers correspond to those indicated in Figure 2.

Peak	RT [min]	Molecular Ion [M-H]−	Ion Mode	MS/MS Fragments	Formula	Error [ppm]	mSigma	Tentative Identification
1	1.4	315.0715	-	315(100), 152(8)	$C_{13}H_{15}O_9$	2.0	0.3	protocatechuoyl—hexoside
2	1.6	153.0187	-	153(100), 109(40)	$C_7H_5O_4$	4.3	14.4	protocatechuic acid
3	2.0	577.1340	-	289(100), 407(72)	$C_{30}H_{25}O_{12}$	2.1	4.2	procyanidin isomer
4	2.3	289.0712	-	289(100)	$C_{15}H_{13}O_6$	1.8	2.3	catechin
5	2.5	353.0867 577.1341	- -	191(100), 353(6) 289(100), 407(71)	$C_{16}H_{17}O_9$ $C_{30}H_{25}O_{12}$	3.1 1.8	23.4 7.4	chlorogenic acid procyanidin isomer
6	2.9	577.1344	-	289(100), 407(72)	$C_{30}H_{25}O_{12}$	1.2	13.7	procyanidin isomer
7	3.3	289.0714	-	289(100), 245(19), 203(9)	$C_{15}H_{13}O_6$	1.3	5.2	catechin
8	4.8	609.1457	-	299(100), 271(63), 447(12)	$C_{27}H_{29}O_{16}$	0.7	17.2	quercetin-hexoside-deoxyhexoside
9	5.2	593.1516	-	283(100), 285(43), 447(36)	$C_{27}H_{29}O_{15}$	−0.7	23.8	kaempferol-hexoside-deoxyhexoside
10	5.4	609.1469	-	271(100), 300(68)	$C_{27}H_{29}O_{16}$	−1.2	10.4	quercetin-hexoside-deoxyhexoside
11	5.5	465.1014 595.1634	+ +	303(100) 287(100)	$C_{21}H_{21}O_{12}$ $C_{27}H_{31}O_{15}$	2.8 3.9	4.9 7.6	quercetin-hexoside kaempferol-hexoside-deoxyhexoside
12	5.7	449.1066 595.1634	+ +	303(100) 303(100), 449(4)	$C_{21}H_{21}O_{11}$ $C_{27}H_{31}O_{15}$	2.7 2.8	3.8 2.4	quercetin-deoxyhexoside quercetin-di-deoxyhexoside
13	6.4	595.1640	+	287(100)	$C_{27}H_{31}O_{15}$	3.0	5.3	kaempferol-hexoside-deoxyhexoside
14	6.6	449.1062 579.1690 711.2102	+ + +	287(100) 287(100), 433(4) 287(100), 433(4)	$C_{21}H_{21}O_{11}$ $C_{27}H_{31}O_{14}$ $C_{32}H_{39}O_{18}$	3.6 3.2 4.1	4.9 6.0 11.5	kaempferol-hexoside kaempferol-di-deoxyhexoside kaempferol-di-deoxyhexoside-pentoside
15	6.7	447.0937	-	271(100), 300(73)	$C_{21}H_{19}O_{11}$	−0.9	3.1	quercetin-deoxyhexoside
16	7.5	463.0886	-	301(100)	$C_{21}H_{19}O_{12}$	−0.8	15.0	quercetin-hexoside
17	9.8	595.1429	+	147(100), 287(34)	$C_{30}H_{27}O_{13}$	2.9	24.8	kaempferol-hexoside-deoxyhexoside-coumaric acid
18	10.3	593.1854	+	285(100), 447(6)	$C_{28}H_{33}O_{14}$	1.8	9.2	unidentified-hexoside-deoxyhexoside (flavonoid)
19	13.0	327.2174	-	327(100), 211(9)	$C_{18}H_{31}O_5$	1.0	1.5	unidentified
20	14.2	329.2327	-	329(100), 211(17)	$C_{18}H_{33}O_5$	2.1	3.2	unidentified
21	14.6	289.2375	+	235(100), 253(74), 217(64), 135(19), 161(14)	$C_{16}H_{33}O_4$	−0.7	8.0	unidentified

Table 8. Compounds identified in the ribwort plantain (*Plantago lanceolata* L.) extract using UHPLC-QTOF-MS/MS. Compound numbers correspond to those indicated in Figure 2.

Peak	RT [min]	Molecular Ion [M-H]$^-$	Ion Mode	MS/MS Fragments	Formula	Error [ppm]	mSigma	Tentative Identification
1	1.4	373.1142	-	373(100), 211(92), 123(24)	$C_{16}H_{21}O_{10}$	−0.4	8.3	geniposidic acid
2	2.1	461.1672	-	461(100), 315(3)	$C_{20}H_{29}O_{12}$	−1.7	7.5	decaffeoylacteoside
3	3.3	451.2192	-	405(100), 179(8), 243(8), 167(7)	$C_{20}H_{35}O_{11}$	−1.7	4.6	caryoptoside isomer
4	3.9	813.1363	-	285(100), 637(23), 351(18), 461(15),	$C_{33}H_{33}O_{24}$	0.6	7.9	luteolin-tri-oxyhexide isomer
5	4.1	637.1044	-	285(100), 351(89)	$C_{27}H_{25}O_{18}$	0.4	20.9	luteolin-di-oxyhexide isomer
6	4.6	639.1941	-	639(100), 621(42), 161(28), 135(9), 447(7)	$C_{29}H_{35}O_{16}$	−1.6	15.8	unidentified phenylethanoid
7	4.7	639.1933	-	639(100), 621(59), 161(28), 133(16), 475(11)	$C_{29}H_{35}O_{16}$	−0.4	11.7	unidentified phenylethanoid
8	5.4	637.1046	-	285(100), 461(44)	$C_{27}H_{25}O_{18}$	0.0	7.6	luteolin-di-oxyhexoside isomer
9	5.7	461.0721 989.1849 639.1935	- - -	285(100) 285(100), 461(12), 813(8) 639(100), 285(64), 161(41), 477(32), 461(21)	$C_{21}H_{17}O_{12}$ $C_{43}H_{41}O_{27}$ $C_{29}H_{35}O_{16}$	1.1 −0.8 −0.7	6.8 43.6 8.1	luteolin-oxyhexoside isomer unidentified luteolin derivative (flavonoid) unidentified phenylethanoid
10	6.3	545.2231 477.1400 755.2407	- - -	545(100), 337(14), 235(10), 193(8) 477(100), 161(22), 133(11), 315(9), 179(2) 755(100), 161(21), 179(10), 593(10), 135(8)	$C_{25}H_{37}O_{13}$ $C_{23}H_{25}O_{11}$ $C_{34}H_{43}O_{19}$	1.5 0.4 −0.4	10.4 9.5 12.5	unidentified calceolarioside A isomer forsythoside isomer
11	6.5	623.1980 639.1930	- -	623(100), 161(26), 461(12) 639(100), 161(30), 477(13), 133(6)	$C_{29}H_{35}O_{15}$ $C_{29}H_{35}O_{16}$	0.2 0.2	5.6 16.6	verbascoside isomer unidentified phenylethanoid
12	6.9	445.0779 755.2401	- -	269(100) 755(100), 161(24), 593(11), 133(8)	$C_{21}H_{17}O_{11}$ $C_{34}H_{43}O_{19}$	−0.6 0.4	24.9 19.4	apigenin-oxyhexoside forsythoside isomer
13	7.1	623.1987	-	623(100), 161(17), 461(7), 133(5)	$C_{29}H_{35}O_{15}$	−0.9	7.0	verbascoside isomer
14	7.5	475.0877	-	274(100), 299(73)	$C_{22}H_{19}O_{12}$	1.0	7.5	kaempferide-oxyhexoside
15	8.0	637.2138	-	637(100), 461(59), 175(41)	$C_{30}H_{37}O_{15}$	0.0	15.9	leucoseptoside A isomer
16	8.0	621.1826		321(100), 323(21), 179(18), 487(14)	$C_{29}H_{33}O_{15}$	−0.2	4.8	unidentified
17	8.4	629.2674	-	583(100), 421(21), 451(13), 289(9)	$C_{26}H_{45}O_{17}$	−1.9	5.9	unidentified
18	10.5	651.2304	-	651(100), 175(24), 160(12), 193(7), 475(5)	$C_{31}H_{39}O_{15}$	−1.5	12.5	unidentified

Table 9. Compounds identified in the marshmallow (*Althaea* L.) extract using UHPLC-QTOF-MS/MS. Compound numbers correspond to those indicated in Figure 2.

Peak	RT [min]	Molecular Ion [M-H]⁻	Ion Mode	MS/MS Fragments	Formula	Error [ppm]	mSigma	Tentative Identification
1	1.4	326.1246	-	164(100), 236(26), 147(20)	$C_{15}H_{20}NO_7$	−0.2	16.1	phenylalanyl-hexoside
2	2.5	417.1043	-	417(100), 152(8)	$C_{17}H_{21}O_{12}$	−1.0	4.6	gentisic acid-dipentoside
3	2.8	179.0357 361.0966	- -	179(100), 135(62) 361(100), 281(13), 171(7)	$C_9H_7O_4$ $C_{24}H_{13}N_2O_2$	−4.1 4.7	41.7 55.9	caffeic acid unidentified
4	3.5	227.0568	-	227(100), 139(22), 165(9)	$C_{10}H_{11}O_6$	−3.1	9.2	unidentified
5	3.7	623.0037	-		$C_{17}H_{19}O_{23}S$	1.0	32.2	flavonoid disulfo-hexoside
6	5.0	193.0479	+	193(100), 134(36), 178(36), 191(10)	$C_{10}H_9O_4$	8.3	9.7	scopoletin isomer
7	5.2	425.0562	-	425(100), 297(88), 315(86), 241(75), 327(73)	$C_{14}H_{17}O_{15}$	2.5	38.1	unidentified
8	5.9	636.9843	-		$C_{17}H_{16}O_{24}S$	−1.7	12.3	flavonoid sulfo-glycoside
9	6.1	733.0950	-		$C_{28}H_{29}O_{21}S$	−3.1	37.4	flavonoid sulfo-glycoside
10	6.3	541.0317	-	254(100), 285(33), 175(9)	$C_{21}H_{17}O_{15}S$	−4.3	12.5	theograndin I isomer
11	6.4	433.1521	-	433(100), 403(86), 311(76), 299(58)	$C_{22}H_{25}O_9$	−3.9	5.5	unidentified
12	6.7	527.0522 639.1223	- -	285(100), 527(4), 213(4), 447(4) 301(100), 371(25), 299(24), 459(22)	$C_{21}H_{19}O_{14}S$ $C_{27}H_{27}O_{18}$	−4.1 −3.1	40.7 62.4	flavonoid sulfo-glycoside unidentified-deoxyhexoside-hexoside (flavonoid)
13	7.2	524.2881	-	524(100), 362(50)	$C_{27}H_{42}NO_9$	−3.1	25.3	unidentified
14	7.3							unidentified
15	7.6	557.0266	-	301(100), 254(77), 315(58), 271(42)	$C_{21}H_{17}O_{16}S$	−4.2	21.0	theograndin II isomer
16	8.5	555.0464	-	254(100), 284(30), 299(24), 175(7)	$C_{22}H_{19}O_{15}S$	−2.5	9.6	flavonoid sulfo-oxyhexoside
17	8.9	541.0672	-	299(100), 284(68), 461(6)	$C_{22}H_{21}O_{14}S$	−2.7	14.1	flavonoid sulfo-hexoside
18	8.9	541.0310	-	285(100), 254(67), 461(22)	$C_{21}H_{17}O_{15}S$	−3.1	34.2	flavonoid sulfo-oxyhexoside
19	9.1	571.0411	-	300(100), 254(79), 315(46), 491(23)	$C_{22}H_{19}O_{16}S$	−2.0	24.2	flavonoid sulfo-oxyhexoside
20	9.4	307.0731	-	233(100), 205(76), 263(75), 191(9)	$C_{17}H_{11}N_2O_4$	−2.2	2.1	unidentified
21	10.5	475.0896	-		$C_{22}H_{19}O_{12}$	−3.1	56.1	unidentified
22	11.4	555.0469	-	254(100), 284(80), 299(71), 475(35), 175(40)	$C_{22}H_{19}O_{15}S$	−3.4	1.8	flavonoid sulfo-oxyhexoside
23	18.1	311.2241	-		$C_{18}H_{31}O_4$	−4.2	4.5	unidentified
24	19.3	459.2037	-		$C_{25}H_{31}O_8$	−2.8	24.7	unidentified
25	19.8	459.2030	-		$C_{25}H_{31}O_8$	−1.3	56.6	unidentified

4. Discussion

4.1. Antimicrobial Activity of Plant Extracts

Our experiments showed that with the increase in dose the tested plant extracts gradually limited the acidifying activity of the tested LAB, but did not limit the viability of these cells. This finding is interesting considering the necessity to keep the LAB cells alive and maintain them biologically active in fermented milk drinks throughout the declared shelf life. Perhaps, these plant extracts could be used in the dose range studied to stop the activity of starter bacteria and consequently prevent acidification of the fermented milk beverages with the cultures used in this study. It should be noted that there are no studies to date in this regard.

Various spices and herbal extract may affect activity or vitality of lactic acid bacteria, and this phenomenon depend on the genus, species and even sometimes strain of lactic bacteria, the species of herbs and the method of obtaining the extract from them. It is known from research that some LABs have the ability to degrade certain phenolic compounds found in food, including those with high antioxidant activity [33]. Polyphenol-resistant bacteria have the ability to metabolize polyphenol compounds, depending on their chemical structure (substitutions in the phenolic ring) and concentration. *Lactobacillus plantarum* has been described to exhibit several enzymatic activities, such as that of tannase, phenolic acid decarboxylase (PAD), and benzyl alcohol dehydrogenase which can degrade some phenolic compounds [33]. Only limited studies have analyzed the influence of phenolic compounds on the growth and viability of other LAB species [34,35].

We did not observe any negative or positive effect of the added extracts on the viability of lactobacilli cells during the fermentation process. This is in line with the results reported by Otaibi and Demerdash [36], who showed that appropriately selected portions of sage extracts resulted in better survival of yogurt bacterial cells, while El-Nawawy et al. [37] indicated a beneficial effect of extracts on the multiplication of bacteria in yogurt. An increase was observed in the numbers of *S. thermophilus* cells than *L. delbrueckii* subsp. *bulgaricus*. However, after digestion, the number of *Streptococcus* cells decreased by an average of 1 log cycle, while the population of *Lactobacillus* cells remained stable. This is also confirmed by the studies of Zaręba et al. [38] and Ziarno and Margol [39], which proves that LAB poorly tolerate digestive juices, but their survival rate is highly dependent on the culture used. The results obtained in this study thus confirm that the viability of LAB in selected starter cultures can be maintained at a fairly high level, even under such drastic conditions as digestive juices.

There are studies on the antimicrobial effect of plant extracts available in the literature. The antimicrobial activity of plant extracts investigated in this work is often the subject of scientific research [40], but only a few concern the effect of extracts on the activity or population of selected LABs [36]. Due to the lack of comprehensive information on the effect of selected plant extracts on LABs, it is worth taking a brief look at the effect of other plant additives on the bacteria in question. Fortunately, a few reports are found in the literature on the beneficial or inhibitory effects of different plant additives on LABs [41–46].

Among the few publications dealing with the effects of valerian (*V. officinalis* L.) on bacteria, there are some reports on its antibacterial activity [47]. However, no data are available regarding the effect of purge on LABs. Sage (*S. officinalis* L.) oils have proven antibacterial and antifungal properties [40,48–50]. Their effect on yogurt bacteria has also been studied proving that the appropriate portions of these additives do not inhibit the growth of yogurt bacteria, and on the contrary, they may have a positive effect on their survival and increase their number during storage [36]. This was also confirmed by the results of our research. Moreover, El-Nawawy et al. [37] reported a beneficial effect of sage extracts on the population of *S. thermophilus* and *L. bulgaricus*. On the other hand, Hołderna-Kedzia and Kedzia [51] showed a negative effect of these extracts on *Lactobacillus acidophilus* ATCC 4356, *Lactobacillus casei* ATCC 393, *Lactobacillus rhamnosus* Hansen 1968, *Bifidobacterium bifidum* ATCC 35914, *S. thermophilus* ATCC 14485, and *Saccharomyces boulardii* SB48 ATCC-MYA-796. The antibacterial effect of chamomile (M. chamomilla L.) extracts has been described in many studies [49,51–53]. However, there are no scientific reports on the effect of chamomile extract on LABs. Several researchers have studied the activity of cistus (*Cistus* L.) extract against pathogenic fungi and bacteria [54–57]. Few reports even indicate that linden blossom has a slight antibacterial effect (*Tilia* L.) [58,59], but there are no data on its influence on LABs. Ethanol and methanol extracts of ribwort plantain (*P. lanceolata* L.) have already been tested against the strains of *Staphylococcus aureus*, *Staphylococcus epidermidis*, *Bacillus subtilis*, *Bacillus cereus*, *Proteus mirabilis*, *Pseudomonas aeruginosa*, *Escherichia coli*, *Enterobacter aerogenes*, *Klebsiella pneumoniae*, *Candida albicans*, and *Candida tropicalis* and found to be active against these bacteria [60–63].

For marshmallow (*Althaea* L.) extracts, it has been shown that the extracts inhibited the growth of bacteria such as *Pseudomonas fluorescens*, *P. aeruginosa*, *Bordetella bronchiseptica*, *S. aureus*, *S. epidermidis*, *Micrococcus luteus*, *Enterococcus faecalis*, *B. subtilis*, *B. cereus*, *Aspergillus niger*, *C. albicans*, and *Saccharomyces cerevisiae* [64,65]. It was also showed that marshmallow extracts did not have any inhibitory effect on *E. coli*, *K. pneumoniae*, or *Serratia marcescens* [64]. However, the effect of these extracts on LABs has not yet been studied.

4.2. TPC and Antioxidant Capacities of Plant Extracts

It is reported that the metabolites present in plant extracts, including phenolic compounds, are responsible for their various biological activities, such as total phenolic content and antioxidant activity [20,22,66–68]. Therefore, it is worth analyzing the research proving the antioxidant activity of selected herbal extracts. Phenolic compounds are widespread in the world of plants. Based on the structure of the carbon skeleton, phenolic compounds can be divided into phenolic acids and flavonoids.

Wang et al. [69] and Şen and Mat [66] indicated that sesquiterpenes, iridoids, flavonoids, and alkaloids were isolated from valerian (*V. officinalis* L.) extracts. Katsarova et al. [67] showed the lowest antioxidant activity for valerian extracts among the eight tested plant extracts (*V. officinalis*, *Melissa officinalis*, *Crataegus monogyna*, *Hypericum perforatum*, *Serratula coronata*, and their two combinations): oxygen radical absorbing capacity—820.5 ± 21.9 μmol TE/g; hydroxyl radical averting capacity—381.6 ± 14.0 μmol GAE/g; and polyphenolic content—43.36 ± 1.3 mg/g. In our study, the valerian extract showed slightly less TPC, expressed as mg GAE/g extract, compared to the above value, despite the fact that its chemical composition included a large diversity of ingredients with antioxidant properties. Furthermore, compared to the other tested plant extracts, its antioxidant capacities were lower.

The crude extract of sage (*S. officinalis* L.) studied by Koşar et al. [70] contained hydroxybenzoic acids, hydroxycinnamic acids, flavonoids, and diterpenoids, in addition to caffeic acid, carnosic acid, luteolin, luteolin-7-O-glucoside, and rosmarinic acid. However, the composition of the plant extract may vary depending on the method used for extraction [5,71–73]. The sage extract obtained by Houghton [5] also contained cyclic monoterpenes, such as 1,8-cineol, α-pinene, and camphor The crude extract and subfractions demonstrated varying degrees of antioxidant capacity. Rosmarinic acid and abietane diterpenes were thought to be responsible for the potent scavenging properties of Salvia taxa [74,75]. Our research confirmed a good correlation between the high TPC in the sage extract and its high antioxidant capacities. Modern analytical methods based on semipreparative HPLC, high-resolution MS, nuclear magnetic resonance spectroscopy, infrared spectroscopy, and single-crystal X-ray diffraction were used to analyze the chemical composition of different sage extracts obtained from the areal parts. A substantial difference was found between the composition of sage flower CO2 extract and that of sage flower resin extract [76]. In addition to the known compounds, novel compounds were identified in sage flowers. Among these, some were preidentified in our research, namely danshensu, caffeic acid, rosmarinic acid, luteolin, and apigenin.

The basic active substances of chamomile (*M. chamomilla* L.) include essential oils, flavonoids, coumarins, sesquiterpenes, polyacetylenes, spiroether, choline, mucus compounds, vitamin C, and mineral salts. Chamomile oil contains compounds that have a specific healing effect, such as antiallergic chamazulene and bisabolol and its oxides. Many bioactive phenolic compounds, including herniarin and umbelliferone (coumarin), chlorogenic acid and caffeic acid (phenylpropanoids), apigenin, apigenin-7-O-glucoside, luteolin and luteolin-7-O-glucoside (flavones), quercetin and rutin (flavonols), and naringenin (flavanone), have been found in chamomile extract [77–79]. Among flavonoids, apigenin is the most promising due to its multiple therapeutic functions. It exists in the form of various glycosides or in very small quantities as free apigenin. The method of extraction not only determines the chemical composition of chamomile extract but also its antioxidant activity [72]. Flavonoids represent the major fraction of water-soluble components in chamomile [71]. Their values in chamomile aqueous extract were established as follows: quercetin equivalent per gram of extract—27.65 ± 0.007 μg; GAE per gram of extract—146.97 ± 0.046 μg, and tannic acid equivalent per gram of extract—132.22 ± 0.023 μg. In contrast, analysis of chamomile extract containing many flavonoids and numerous organic acids and their derivatives in this study revealed the low TPC compared to other tested plant extracts tested, as well as one of the lowest antioxidant capacities.

Viapiana et al. [9] determined the content of phenolic acids and flavonoids in cistus (*Cistus* L.) extracts. Their results revealed that aqueous cistus extracts were richer in phenolic compounds and showed strong antioxidant activities. The total amount of polyphenols in the leaf, stalk, and bud extracts of *Cistus incanus* determined by Dimcheva and Karsheva [73] varied between 36.26 and 115.32 mg GAE/g dry weight (dw), depending on the time of extraction. After 180 min of *C. incanus* extraction, the phenolic content was slightly higher than that observed in our study. Such result has also been shown by other studies. Researchers showed that the place of origin is the main factor differentiating the antibacterial activities of cistus samples. Dimcheva et al. [10] found catechins, flavonoids, and gallic and vanillic acids in *Cistus* extracts. The compounds identified by Gori et al. [80] in crude ethanolic leaf extract of *C. incanus* included gallic acid derivatives, condensed tannins, and flavonol glycosides. In total, 19 compounds were identified based on the fragmentation of individual polyphenols and by comparing their retention times and UV/Vis spectra with authentic standards. As in our research, the presence of epicatechin, myricetin hexoside, and quercetin pentoside was found in the C. incanus extract by Gori et al. [80]. In our study, the *Cistus* extract had the highest TPC among the tested plant extracts, and thus showed the highest antioxidant capacities. Its composition included many organic acids and their derivatives, as well as flavonoids such as catechin, quercetin, kaempferol, and myricetin.

Several studies have been performed on linden blossom (*Tilia* L.) extracts. The results revealed the presence of terpenoids, quercetin, and kaempferol derivatives (such as tiliroside), phenolic compounds, esters, aliphatic acids, hydrocarbons, condensed tannins, and a coumarin scopoletin in the extracts [11,16,81,82]. Wissam et al. [83] stated that TPC and the antioxidant activity were determined in the ethanolic extracts of dried linden blossom leaves (0.3303 ± 0.0896 mg/mL calculated as DPPH scavenging activity). In our study, the DPPH value of linden blossom extract was determined as 0.161 ± 0.007, but expressed as mmol TE/g extract. This value was the second highest recorded in our experiments. TPC of the linden blossom extract was also one of the largest and resulted from the presence of quercetin, kaempferol, other flavonoids, and their derivatives, as well as many organic acids. In addition, 3,4-dihydroxybenzoic acid, myricetin, rutin, ferulic acid, and 3,4-dihydroxybenzaldehyde were found in abundance in the Tilia tomentosa flower [84]. In turn, the major phenolic compound observed in acetone and methanolic extracts of Tilia argentea was protocatechuic acid. The leaf samples of Tilia species were found to contain many compounds similar to those in flower samples, but each of these samples possessed a unique chemical profile including the percentage and type of flavonoid constituents [85].

Some scientists showed that phenolic compounds, mainly flavonoids and hydroxycinnamic acids, were the main components of hydrophilic ribwort plantain (*P. lanceolata* L.) extracts [19]. Galvez et al. [20] found that luteolin was biologically important among the flavonoids. According to [17], phenylethanoids, especially Aukubin, are responsible for the antimicrobial effects of ribwort plantain extracts. It was reported that these extracts exhibited a strong antioxidant activity [19]. Lukova et al. [68] studied the antioxidant activity of the ethanol extracts of *P. lanceolata* leaves by DPPH scavenging test, CUPRAC (cupric reducing antioxidant capacity) assay, and FRAP assay and established the following values: DPPH—59.04 ± 0.09%; CUPRAC—21.9 ± 0.58 μM TE/g dw; and FRAP—51.85 ± 1.54 μM TE/g dw. The Plantago plantain leaf extract tested in our study was characterized by an average content of TPC expressed as mg GAE/g of extract, as well as average antioxidant capacities expressed by DPPH, ABTS, or FRAP scavenging ability.

In general, marshmallow (*Althaea* L.) ethanol extracts show high antioxidant activity, which is due to the presence of active compounds such as flavonoids and mucus polysaccharides [22]. However, this was not confirmed by the results of our research. The leaves of marshmallow contain the coumarin scopoletin, as well as many flavonoids (hypolaetin-8-glucoside, isoquercitrin, kaempferol, caffeic acid, p-coumaric acid, ferulic acid, p-hydroxybenzoic acid, salicylic acid, p-hydroxyphenylacetic acid, vanillic acid) [86,87]. Elmastas et al. [22] indicated the strong total antioxidant activity of ethanolic marshmallow extract. They reported that the marshmallow extract showed effective reducing power, free radical scavenging activity, superoxide anion radical scavenging activity, and metal chelating ability at the same concentration (50, 100, and 250 mg/mL, respectively). In comparison, the marshmallow extract tested in our study contained only a few antioxidant substances (some flavonoids and organic acids), which resulted in its lower antioxidant capacities.

5. Conclusions

This study showed that the addition of herbal extracts from valerian (*V. officinalis* L.), sage (*S. officinalis* L.), chamomile (*M. chamomilla* L.), cistus (*Cistus* L.), linden blossom (*Tilia* L.), ribwort plantain (*P. lanceolata* L.), and marshmallow (*Althaea* L.) did not inhibit the growth of LAB in fermented milk drinks such as yogurts. In light of the presented results, yogurts enriched with the plant extracts tested in this study can be of interest to customers. However, these herbal extracts should be added in limited amounts because they gradually inhibit the fermentation activity. Now, knowing in what dose range dairy products fermented with the addition of selected herbal extracts can be tested in the future, including storage research. When added at concentrations above 2% by weight, which probably can be used to prevent overacidification of fermented milk after the fermentation process is complete, herbal extracts from valerian, sage, chamomile, cistus, linden blossom, ribwort plantain, or marshmallow should be tested for the storage stability of fermented milk beverages such as yoghurts containing live lactic acid bacteria.

Nevertheless, due to the antioxidant properties, a slight addition of the herbal extracts containing polyphenols to different kinds of food products can increase the nutritional quality, thus making them functional foods. The microflora of yogurts containing the tested herbal extracts did not die during digestion in model digestive juices, and this amount of bacteria surviving digestion can benefit the health of consumers. Thus, the tested plant extracts had neither an inhibitory nor a stimulating effect on bacteria in the fermented milk samples.

Author Contributions: Conceptualization, M.Z. and M.K. (Mariola Kozłowska); methodology, M.Z., M.K. (Mariola Kozłowska) and I.Ś.; investigation, M.Z., M.K. (Mariola Kozłowska), I.Ś., M.K. (Mariusz Kowalczyk), S.P., A.S. and B.S.; data curation, M.Z. and B.S.; writing—M.Z., M.K. (Mariola Kozłowska), I.Ś. and B.S.; writing—review and editing, M.Z., M.K. (Mariola Kozłowska), I.Ś. and S.P.; project administration, M.Z. All authors have read and agreed to the published version of the manuscript.

Funding: This work was supported by Polish Ministry of Science and Higher Education with funds of the Warsaw University of Life Sciences WULS–SGGW (Poland).

Institutional Review Board Statement: This study did not involve humans or animals.

Informed Consent Statement: This study did not involve humans.

Data Availability Statement: Data sharing not applicable.

Conflicts of Interest: The authors declare no conflict of interest.

References

1. Goun, E.A.; Petrichenko, V.M.; Solodnikov, S.U.; Suhinina, T.V.; Kline, M.A.; Cunningham, G.; Miles, H. Anticancer and antithrombin activity of Russian plants. *J. Ethnopharmacol.* **2002**, *81*, 337–342. [CrossRef]
2. O'Hara, M.; Kiefer, D.; Farrell, K.; Kemper, K. A review of 12 commonly used medicinal herbs. *Arch. Fam. Med.* **1998**, *7*, 523–536. [CrossRef]
3. Woźniak, M.; Ostrowska, K.; Szymański, Ł.; Wybieralska, K.; Zieliński, R. Aktywność przeciwrodnikowa ekstraktów szałwii i rozmarynu. *Żywność* **2009**, *4*, 133–141, (Abstract in English).
4. Tildesley, N.T.; Kennedy, D.O.; Perry, E.K.; Ballard, C.G.; Savelev, S.A.W.K.; Wesnes, K.A.; Scholey, A.B. *Salvia lavandulaefolia* (Spanish Sage) enhances memory in health young volunteers. *Pharmacol. Biochem. Behav.* **2003**, *75*, 669–674. [CrossRef]
5. Houghton, P.J. Activity and constituents of sage relevant to the potential treatment of symptoms Alzheimer's disease. *Herba Gran* **2004**, *61*, 38–43.
6. Khayyal, M.T.; El-Ghazaly, M.A.; Kenawy, S.A.; Seif-El-Nasr, M.; Mahran, L.G.; Kafafi, Y.A.; Okpanyi, S.N. Antiulcerogenic effect of certain plant extracts and their combinations. *Arzneimittelforsch* **2001**, *51*, 545–553.
7. Torrado, S.; Agis, A.; Jimenez, M.E.; Cadorniga, R. Effect of dissolution profile and (−)-alpha-bisabolol on the gastrotoxicity of acetylsalicylic acid. *Pharmazie* **1995**, *50*, 141–143. [PubMed]
8. Sebai, H.; Jabri, M.A.; Souli, A.; Rtibi, K.; Selmi, S.; Tebourbi, O.; Sakly, M. Antidiarrheal and antioxidant activities of chamomile (*Matricaria recutita* L.) decoction extract in rats. *J. Ethnopharmacol.* **2014**, *152*, 327–332. [CrossRef]
9. Viapiana, A.; Konopacka, A.; Waleron, K.; Wesolowski, M. *Cistus incanus* L. commercial products as a good source of polyphenols in human diet. *Ind. Crops Prod.* **2017**, *107*, 297–304. [CrossRef]
10. Dimcheva, V.; Kaloyanov, N.; Karsheva, M. The polyphenol composition of *Cistus incanus* L., *Trachystemon orientalis* L. and *Melissa officinalis* L. infusions by HPLC-DAD method. *Open J. Anal. Bioanal. Chem.* **2019**, *3*, 031–038. [CrossRef]
11. Arcos, M.L.B.; Cremaschi, G.; Werner, S.; Coussio, J.; Ferraro, G.; Anesini, C. *Tilia cordata* Mill. extracts and scopoletin (isolated compound): Differential cell growth effects on lymphocytes. *Phytother. Res.* **2006**, *20*, 34–40. [CrossRef]
12. Hyun Kim, K.; Moon, E.; Kim, S.Y.; Choi, S.U.; Lee, K.R. Lignan constituents of Tilia amurensis and their biological evaluation on antitumor and anti-inflammatory activities. *Food. Chem. Toxicol.* **2012**, *50*, 3680–3686.
13. Martinez, A.L.; Gonzalez-Trujano, M.A.; Aguirre-Hernandez, E.; Moreno, J.; Soto-Hernandez, M.; Lopez-Munoz, F.J. Antinociceptive activity of *Tilia americana* var. mexicana inflorescences and quercetin in the formalin test and in an arthritic pain model in rats. *Neuropharmacology* **2009**, *56*, 564–571. [CrossRef]
14. Toker, G.; Küpeli, E.; Memisoğlu, M.; Yesilada, E. Flavonoids with antinociceptive and anti-inflammatory activities from the leaves of *Tilia argentea* (silver linden). *J. Ethnopharmacol.* **2004**, *95*, 393–397. [CrossRef]
15. Aguirre-Hernandez, E.; Martinez, A.L.; Gonzalez-Trujano, M.A.; Moreno, J.; Vibrans, H.; Soto-Hernandez, M. Pharmacological evaluation of the anxiolytic and sedative effects of *Tilia americana* L. var. mexicana in mice. *J. Ethnopharmacol.* **2007**, *109*, 140–145. [CrossRef]
16. Herrera-Ruiz, M.; Román-Ramos, R.; Zamilpa, A.; Tortoriello, J.; Jiménez-Ferrer, J.E. Flavonoids from *Tilia americana* with anxiolytic activity in plus-maze test. *J. Ethnopharmacol.* **2008**, *118*, 312–317. [CrossRef] [PubMed]
17. ESCOP. *Monographs 2013: European Scientific Cooperative on Phytotherapy. Plantaginis lanceolatae Folium/Herba*, 2nd ed.; Ribwort Plantain Leaf/Herb; European Scientific Cooperative on Phytotherapy (ESCOP): Exeter, UK, 2013; pp. 383–387.
18. Parus, A.; Grys, A. *Plantago lanceolata* L.-medicinal properties. *Post Fit.* **2010**, *3*, 162–165, (Abstract in English).
19. Dalar, A.; Türker, M.; Konczak, I. Antioxidant capacity and phenolic constituents of *Malva neglecta* Wallr. and *Plantago lanceolata* L. from Eastern Anatolia Region of Turkey. *J. Herbal Med.* **2012**, *2*, 42–51. [CrossRef]
20. Galvez, M.; Martí, C.; Lopez-Lazaro, M.; Cortes, F.; Ayuso, J. Cytotoxic effect of Plantago spp. on cancer cell lines. *J. Ethnopharmacol.* **2003**, *88*, 125–130. [CrossRef]
21. Deters, A.; Zippel, J.; Hellenbrand, N.; Pappai, D.; Possemeyer, C.; Hensel, A. Aqueous extracts and polysaccharides from Marshmallow roots (*Althea officinalis* L.): Cellular internalisation and stimulation of cell physiology of human epithelial cells in vitro. *J. Ethnopharmacol.* **2010**, *127*, 62–69. [CrossRef] [PubMed]
22. Elmastas, M.; Ozturk, L.; Gokce, I.; Erenler, R.; Aboul Enein, H.Y. Determination of antioxidant activity of marshmallow flower (*Althaea officinalis* L.). *Anal. Lett.* **2004**, *37*, 1859–1869. [CrossRef]
23. Izzo, A.A.; Di Carlo, G.; Mascolo, N.; Autore, G.; Capasso, F. Antiulcer effect of flavonoids. Role of endogenous PAF. *Phytother. Res.* **1994**, *29*, 87–93. [CrossRef]
24. Guerrero, J.A.; Lozano, M.L.; Castillo, J.; Benavente Garcia, O.; Vicente, V.; Rivera, J. Flavonoids inhibit platelet function through binding to the thromboxane A2 receptor. *J. Thromb. Haemost.* **2005**, *3*, 369–376. [CrossRef] [PubMed]
25. Kozłowska, M.; Ścibisz, I.; Zaręba, D.; Ziarno, M. Antioxidant properties and effect on lactic acid bacterial growth of spice extracts. *CyTA* **2015**, *13*, 573–577. [CrossRef]
26. Kozłowska, M.; Ziarno, M.; Rudzińska, M.; Tarnowska, K.; Majewska, E.; Kowalska, D. Skład chemiczny olejku eterycznego z kolendry i jego wpływ na wzrost wybranych szczepów bakterii kwasu mlekowego. *Żywność* **2018**, *25*, 97–111. (Abstract in English) [CrossRef]

27. Ziarno, M.; Zaręba, D. Effects of milk components and food additives on survival of three bifidobacteria strains in fermented milk under simulated gastrointestinal tract conditions. *Microb. Ecol. Health Dis.* **2015**, *26*. [CrossRef] [PubMed]
28. Singleton, V.L.; Rossi, J.A. Colorimetry of total phenolics with phosphomolybdic-phodphotonegstics acid reagents. *Am. J. Enol. Viticult.* **1965**, *16*, 144–158.
29. Yen, G.C.; Chen, H.Y. Antioxidant activity of various tea extracts in relation to their antimutagenicity. *J. Agric. Food Chem.* **1995**, *43*, 27–32. [CrossRef]
30. Re, R.; Pellegrini, N.; Proteggente, A.; Pannala, A.; Yang, M.; Rice-Evans, C. Antioxidant activity applying an improved ABTS radical cation decolorization assay. *Free Rad. Biol. Med.* **1999**, *26*, 1231–1237. [CrossRef]
31. Benzie, I.F.; Strain, J.J. The ferric reducing ability of plasma (FRAP) as a measure of "antioxidant power": The FRAP assay. *Anal. Biochem.* **1996**, *239*, 70–76. [CrossRef]
32. Dührkop, K.; Fleischauer, M.; Ludwig, M.; Aksenov, A.A.; Melnik, A.V.; Meusel, M.; Dorrestein, P.C.; Rousu, J.; Böcker, S. SIRIUS 4: A rapid tool for turning tandem mass spectra into metabolite structure information. *Nat. Methods* **2019**, *16*, 299–302. [CrossRef]
33. Rodríguez, H.; Curiel, J.A.; Landete, J.M.; de las Rivas, B.; de Felipe, F.L.; Gómez-Cordovés, C.; Mancheño, J.M.; Muñoz, R. Food phenolics and lactic acid bacteria. *Int. J. Food Microbiol.* **2009**, *132*, 79–90. [CrossRef] [PubMed]
34. Landete, J.M.; Rodríguez, H.; Curiel, J.A.; de las Rivas, B.; Mancheño, J.M.; Muñoz, R. Gene cloning, expresión, and characterization of phenolic acid decarboxylase from *Lactobacillus brevis* RM84. *J. Ind. Microbiol. Biotechnol.* **2010**, *37*, 617–624. [CrossRef]
35. Takagaki, A.; Nanjo, F. Metabolism of (−)-epigallocatechin gallate by rat intestinal flora. *J. Agric. Food Chem.* **2010**, *58*, 1313e1321. [CrossRef]
36. Otaibi, M.A.; Demerdash, H.E. Improvement of the quality and shelf life of concentrated yoghurt (labneh) by the addition of some essential oils. *Afr. J. Microbiol. Res.* **2008**, *2*, 156–161.
37. El-Nawawy, M.A.; El-Kenany, Y.M.; Abd El-Ghaffar, E.A. Effect of some herb plants on the use of yoghurt culture. *Ann. Agric. Sci.* **1998**, *7*, 15–17.
38. Zaręba, D.; Ziarno, M.; Strzelczyk, B. Przeżywalność bakterii fermentacji mlekowej w warunkach modelowych jelita cienkiego. *Żywność* **2008**, *15*, 197–205, (Abstract in English).
39. Ziarno, M.; Margol, B. Research into the Ability of Some Selected Starter Lactic Acid Bacteria (Slab) to Survive in a Model Gastric Juice and Cholesterol Binding Under Those These Conditions. *Żywność* **2007**, *6*, 304–314. (Abstract in English).
40. Al-Turki, A.I.; El-Ziney, M.G.; Abdel-Salam, A.M. Chemical and anti-bacterial characterization of aqueous extracts of oregano, marjoram, sage and licorice and their application in milk and labneh. *J. Food Agric. Environ.* **2008**, *6*, 39–44.
41. Nes, I.F.; Skjelkvale, R. Effect of natural spices and oleoresins on *Lactobacillus plantarum* in the fermentation of dry sausage. *J. Food Sci.* **1982**, *47*, 1618–1625. [CrossRef]
42. Kivanç, M.; Akgül, A.; Doğan, A. Inhibitory and stimulatory effects of cumin, oregano and their essential oils on growth and acid production of *Lactobacillus plantarum* and *Leuconostoc mesenteroides*. *Int. J. Food Microbiol.* **1991**, *13*, 81–85. [CrossRef]
43. Michael, M.; Phebus, R.K.; Schmidt, K.A. Impact of a plant extract on the viability of *Lactobacillus delbrueckii* ssp. *bulgaricus* and *Streptococcus thermophilus* in nonfat yogurt. *Int. Dairy J.* **2010**, *20*, 665–672. [CrossRef]
44. Yangmei, Q.; Jiang, W.; Hou, H.; Zhang, H. Application of Plant Green Juice Powder in Maintaining Stability of Viable Count of Yoghourt in Shelf Life. Patent No. CN 102106386 B, 22 March 1991.
45. China, R.; Mukherjee, S.; Sen, S.; Bose, S.; Datta, S.; Koley, H.; Dhar, P. Antimicrobial activity of Sesbania grandiflora flower polyphenol extracts on some pathogenic bacteria and growth stimulatory effect on the probiotic organism *Lactobacillus acidophilus*. *Microbiol. Res.* **2012**, *167*, 500–506. [CrossRef] [PubMed]
46. Dimitrova, M.; Ivanov, G.; Mihalev, K.; Slavchev, A.; Ivanova, I.; Vlaseva, R. Investigation of antimicrobial activity of polyphenol-enriched extracts against probiotic lactic acid bacteria. *Food Sci. Appl. Biotechnol.* **2019**, *2*. 67–73. [CrossRef]
47. Letchamo, W.; Ward, W.; Heard, B.; Heard, D. Essential oil of Valeriana officinalis L. cultivars and their antimicrobial activity as influenced by harvesting time under commercial organic cultivation. *J. Agric. Food Chem.* **2004**, *52*, 3915–3919. [CrossRef] [PubMed]
48. Hać-Szymańczuk, E.; Lipińska, E.; Błażejak, S.; Bieniak, K. Ocena aktywności przeciwbakteryjnej szałwii lekarskiej (*Salvia officinalis* L.). *Brom. Chem. Toksykol.* **2011**, *44*, 667–672, (Abstract in English).
49. Kędzia, A.; Dera-Tomaszewska, B.; Ziółkowska-Klinkosz, M.; Kędzia, A.W.; Wojtaszek-Słomińska, A.; Czernecka, B. Działanie olejku szałwiowego (*Oleum Salviae lavandulaefoliae*) na bakterie tlenowe izolowane z jamy ustnej, dróg oddechowych i przewodu pokarmowego. *Post. Fitoter.* **2011**, *1*, 238–242, (Abstract in English).
50. Gniewosz, M.; Kraśniewska, K.; Węglarz, Z.; Przybył, J.L. Porównanie przeciwdrobnoustrojowej aktywności etanolowego i wodnego ekstraktu z szałwii lekarskiej (*Salvia officinalis* L.). *Brom. Chem. Toksykol.* **2012**, *45*, 743?748, (Abstract in English).
51. Hołderna-Kędzia, E.; Kędzia, B. Działanie preparatów pochodzenia roślinnego na drobnoustroje probiotyczne. *Post. Fitoter.* **2012**, *1*, 72–77, (Abstract in English).
52. Grys, A.; Kania, M.; Baraniak, J. Rumianek—Pospolita roślina zielarska o różnorodnych właściwościach biologicznych i leczniczych. *Post. Fitoter.* **2014**, *1*, 90–93, (Abstract in English).
53. Roby, M.H.H.; Sarhan, M.A.; Selim, K.A.H.; Khalel, K.I. Antioxidant and antimicrobial activities of essential oil and extracts of fennel (*Foeniculum vulgare* L.) and chamomile (*Matricaria chamomilla* L.). *Ind. Crops Prod.* **2013**, *44*, 437–445. [CrossRef]

54. Ustün, O.; Ozçelik, B.; Akyön, Y.; Abbasoglu, U.; Yesilada, E. Flavonoids with anti-*Helicobacter pylori* activity from *Cistus laurifolius* leaves. *J. Ethnopharmacol.* **2006**, *108*, 457–461. [CrossRef]
55. Hutschenreuther, A.; Birkemeyer, C.; Grötzinger, K.; Straubinger, R.K.; Rauwald, H.W. Growth inhibiting activity of volatile oil from *Cistus creticus* L. against *Borrelia burgdorferi* in vitro. *Pharmazie* **2010**, *65*, 290–295. [PubMed]
56. Tomás, M.L.; Morales-Soto, A.; Barrajón-Catalán, E.; Roldán-Segura, C.; Segura-Carretero, A.; Micol, V. Correlation between the antibacterial activity and the composition of extracts derived from various Spanish Cistus species. *Food. Chem. Toxicol.* **2013**, *55*, 313–322. [CrossRef]
57. Rebaya, A.; Belghith, S.I.; Hammrouni, S.; Maaroufi, A.; Ayadi, M.T.; Chérif, J.K. Antibacterial and Antifungal Activities of Ethanol Extracts of *Halimium halimifolium*, *Cistus salviifolius* and *Cistus monspeliensis*. *Int. J. Pharm. Clin. Res.* **2016**, *8*, 243–247.
58. Gönül, Ş.E.A.; Karapinar, M. Inhibitory effect of linden flower (Tilia flower) on the growth of foodborne pathogens. *Food Microbiol.* **1987**, *4*, 97–100. [CrossRef]
59. Yıldırım, A.; Mavi, A.; Oktay, M.; Kara, A.A.; Algur, Ö.F.; Bilaloğlu, V. Comparison of antioxidant and antimicrobial activities of Tilia (*Tilia argentea* Desf ex DC), sage (*Salvia triloba* L.), and Black tea (*Camellia sinensis*) extracts. *J. Agric. Food Chem.* **2000**, *48*, 5030–5034. [CrossRef]
60. Sharifa, A.A.; Neoh, Y.L.; Iswadi, M.I.; Khairul, O.; Abdul Halim, M.; Jamaludin, M.; Hing, H.L. Effects of methanol, ethanol and aqueous extract of Plantago major on gram positive bacteria, gram negative bacteria and yeast. *Ann. Micr.* **2008**, *8*, 42–44.
61. Kahyaoğlu, M.; Türkoğlu, İ. Antimicrobial activities of some plants collected in Elazığ region. *Dumlupinar Univ. Bilim. Enst. Derg.* **2008**, *15*, 1–8.
62. Ferrazzano, G.F.; Cantile, T.; Roberto, L.; Ingenito, A.; Catania, M.R.; Roscetto, E.; Pollio, A. Determination of the in vitro and in vivo antimicrobial activity on salivary Streptococci and Lactobacilli and chemical characterisation of the phenolic content of a *Plantago lanceolata* infusion. *BioMed Res. Int.* **2015**, *2015*, 286817. [CrossRef]
63. Metiner, K.; Ozkan, O.; Ak, S. Antibacterial effects of ethanol and acetone extract of *Plantago major* L. on Gram positive and Gram negative bacteria. *Kafkas Univ. Vet. Fak. Derg.* **2012**, *18*, 503–505.
64. Bonjar, S. Evaluation of antibacterial properties of some medicinal plants used in Iran. *J. Ethnopharmacol.* **2004**, *94*, 301–305. [CrossRef]
65. Valiei, M.; Shafaghat, A.; Salimi, F. Chemical composition and antimicrobial activity of the flower and root hexane extracts of *Althaea officinalis* in Northwest Iran. *J. Med. Plants Res.* **2011**, *5*, 6972–6976.
66. Şen, B.; Mat, A. Chemical and medicinal evaluations of the Valeriana species in Turkey. *J. Fac. Pharm. İstanbul Üniv.* **2015**, *45*, 267–276.
67. Katsarova, M.; Dimitrova, S.; Lukanov, L.; Sadakov, F.; Denev, F.; Plotnikov, E.; Kandilarov, I.; Kostadinova, I. Antioxidant activity and nontoxicity of extracts from *Valeriana officinalis*, *Melissa officinalis*, *Crataegus monogyna*, *Hypericum perforatum*, Serratula coronata and combinations Antistress 1 and Antistress 2. *Bulg. Chem. Commun.* **2017**, *49*, 93–98.
68. Lukova, P.; Karcheva-Bahchevanska, D.; Dimitrova-Dyulgerova, I.; Katsarov, P.; Mladenov, R.; Iliev, I.; Nikolova, M. A comparative pharmacognostic study and assessment of antioxidant capacity of three species from Plantago genus. *Farmacia* **2018**, *66*, 609–614. [CrossRef]
69. Wang PCh Ran, X.H.; Luo, R.H.; Ma, Q.Y.; Liu, Y.Q.; Zhoum, J.; Zhao, Y.X. Phenolic Compounds from the Roots of *Valeriana officinalis* var. *latifolia*. *J. Braz. Chem. Soc.* **2013**, *24*, 1544–1548. [CrossRef]
70. Koşar, M.; Dorman, H.J.D.; Hüsnü Can Başer, K.; Hiltunen, R. *Salvia officinalis* L.: Composition and Antioxidant-related Activities of a Crude Extract and Selected Sub-fractions. *Nat. Prod. Commun.* **2010**, *9*, 1453–1456. [CrossRef]
71. Seddik, K.; Dalila, B.; Saliha, D.; Saliha, D.; Smain, A.; Noureddine, C.; Abderahmane, B.; Daoud, H.; Lekhmici, A. Polyphenols and antioxidant properties of extracts from *Mentha pulegium* L. and *Matricaria chamomilla* L. *Pharm. Commun.* **2013**, *3*, 35–40. [CrossRef]
72. Ranjbar, A.; Mohsenzade, F.; Chehregani, A.; Khajavi, F.; Sifpanahi, H. Antioxidant capacity of various extracts of *Matricaria Chamomilla*, L. parts. *Compl. Med. J.* **2015**, *4*, 1022–1027.
73. Dimcheva, V.; Karsheva, M. *Cistus incanus* from Strandja Mountain as a source of bioactive antioxidants. *Plants* **2018**, *7*, 8. [CrossRef] [PubMed]
74. Lu, Y.; Foo, L.Y. Antioxidant activities of polyphenols from sage (*Salvia officinalis*). *Food Chem.* **2001**, *75*, 197–202. [CrossRef]
75. Koşar, M.; Dorman, H.J.D.; Baser, K.H.C.; Hiltunen, R. An improved HPLC post-column methodology for the identification of free radical scavenging phytochemicals in complex mixtures. *Chromatographia* **2004**, *60*, 635–638. [CrossRef]
76. Gericke, S.; Lübken, T.; Wolf, D.; Kaiser, M.; Hannig, C.; Speer, K. Identification of new compounds from sage flowers (*Salvia officinalis* L.) as markers for quality control and the influence of the manufacturing technology on the chemical composition and antibacterial activity of sage flower extracts. *J. Agric. Food Chem.* **2018**, *66*, 1843–1853. [CrossRef]
77. Redaelli, C.; Formentini, L.; Santaniello, E. PLC determination of coumarins in *Matricaria chamomilla*. *Planta Med.* **1981**, *43*, 412–413. [CrossRef]
78. Ohe, C.; Miami, M.; Hasegawa, C.; Ashida, K.; Sugino, M.; Kanamori, H. Seasonal variation in the production of the head and accumulation of glycosides in the head of *Matricaria chamomilla*. *Acta Hortic.* **1995**, *390*, 75–82. [CrossRef]
79. Gupta, V.; Mittal, P.; Bansal, P.; Khokra, S.L.; Kaushik, D. Pharmacological potential of *Matricaria recutita*—A review. *Int J. Pharm. Sci. Drug Res.* **2010**, *2*, 12–16.

80. Gori, A.; Ferrini, F.; Marzano, M.A.; Tattini, M.; Centritto, M.; Baratto, M.C.; Pogni, R.; Brunetti, C. Characterisation and antioxidant activity of crude extract and polyphenolic rich fractions from *C. incanus* leaves. *Int. J. Mol Sci.* **2016**, *17*, 1344. [CrossRef] [PubMed]
81. Pietta, P.; Facino, R.M.; Carini, M.; Mauri, P. Thermospray liquid chromatography mass spectrometry of flavonol glycosides from medicinal plants. *J. Chrom. A* **1994**, *661*, 121–126. [CrossRef]
82. Behrens, A.; Maie, N.; Knicker, H.; Kogel-Knabner, I. MALDI-TOF mass spectrometry and PSD fragmentation as means for the analysis of condensed tannins in plant leaves and needles. *Phytochemistry* **2003**, *62*, 1159–1170. [CrossRef]
83. Wissam, Z.; Nour, A.A.; Bushra, J.; Zein, N.; Saleh, D. Extracting and studying the antioxidant capacity of polyphenols in dry linden leaves (*Tilia cordata*). *J. Pharm. Phytochem.* **2017**, *6*, 258–262.
84. Kivrak, Ş.; Göktürk, T.; Kivrak, I. Determination of phenolic composition of *Tilia Tomentosa* flowers using UPLC-ESI-MS/MS. *Int. J. Sec. Metab.* **2017**, *4*, 249–256. [CrossRef]
85. Ashour, M.A.-G. Practical authentication and quality control of *Tilia flos*. *Int. J. Biol. Pharm. Appl. Sci.* **2017**, *6*, 1122–1156.
86. Gudej, J. Flavonoids, phenolic acids and coumarins from the roots of *Althaea officinalis*. *Planta Med.* **1991**, *57*. 284–285. [CrossRef] [PubMed]
87. Shah, S.M.A.; Akhtar, N.; Akram, M.; Shah, P.A.; Saeed, T.; Ahmed, K.; Asif, H.M. Pharmacological activity of *Althaea officinalis* L. *J. Med. Plants Res.* **2011**, *5*, 5662–5666.

Article

Antimicrobial Properties, Cytotoxic Effects, and Fatty Acids Composition of Vegetable Oils from Purslane, Linseed, Luffa, and Pumpkin Seeds

Spyridon A. Petropoulos [1,*], Ângela Fernandes [2], Ricardo C. Calhelha [2], Youssef Rouphael [3], Jovana Petrović [4], Marina Soković [4], Isabel C. F. R. Ferreira [2] and Lillian Barros [2,*]

1 Laboratory of Vegetable Production, Department of Agriculture, Crop Production and Rural Environment, University of Thessaly, Fytokou Street, N. Ionia, 384 46 Volos, Greece
2 Centro de Investigação de Montanha (CIMO), Instituto Politécnico de Bragança, Campus de Santa Apolónia, 5300-253 Bragança, Portugal; afeitor@ipb.pt (Â.F.); calhelha@ipb.pt (R.C.C.); iferreira@ipb.pt (I.C.F.R.F.)
3 Department of Agricultural Sciences, University of Naples Federico II, Via Universita 100, 80055 Portici, Italy; youssef.rouphael@unina.it
4 Institute for Biological Research "Siniša Stanković"—National Institute of Republic of Serbia, University of Belgrade, Bulevar Despota Stefana 142, 11000 Belgrade, Serbia; jovana0303@ibiss.bg.ac.rs (J.P.); mris@ibiss.bg.ac.rs (M.S.)
* Correspondence: spetropoulos@uth.gr (S.A.P.); lillian@ipb.pt (L.B.); Tel.: +30-242-109-3196 (S.A.P.); +351-273-330-901 (L.B.)

Citation: Petropoulos, S.A.; Fernandes, Â.; Calhelha, R.C.; Rouphael, Y.; Petrović, J.; Soković, M.; Ferreira, I.C.F.R.; Barros, L. Antimicrobial Properties, Cytotoxic Effects, and Fatty Acids Composition of Vegetable Oils from Purslane, Linseed, Luffa, and Pumpkin Seeds. *Appl. Sci.* **2021**, *11*, 5738. https://doi.org/10.3390/app11125738

Academic Editors: Gwiazdowska Daniela, Krzysztof Juś and Katarzyna Marchwińska

Received: 15 May 2021
Accepted: 18 June 2021
Published: 21 June 2021

Featured Application: Vegetable oils are a rich source of fatty acids and bioactive compounds with numerous beneficial effects to human health. The presented results showed that seed oils of linseed, purslane, luffa, and pumpkin have significant antimicrobial properties that could find application in the food industry as functional ingredients or as non-synthetic antimicrobial agents in the design of new healthy food products. Moreover, they could be used in mixtures with other oils to design new vegetable oils with functional properties and enhance content in omega-3 fatty acids.

Abstract: In the present study, the antimicrobial and cytotoxic activities, as well as the fatty acids composition in vegetable seed oils from linseed, purslane, luffa, and pumpkin were evaluated. For this purpose, two linseed oils and one luffa oil were commercially obtained, while purslane and pumpkin oils were obtained from own cultivated seeds. The results showed a variable fatty acids composition among the tested oils, with α-linolenic, linoleic, oleic, palmitic, and stearic acid being the most abundant compounds. In regards to particular oils, linseed oils were a rich source of α-linolenic acid, luffa and pumpkin oil were abundant in linoleic acid, while purslane oil presented a balanced composition with an almost similar amount of both fatty acids. Luffa oil was the most effective against two of the tested cancer cell lines, namely HeLa (cervical carcinoma) and NCI-H460 (non-small cell lung cancer), while it also showed moderate toxicity against non-tumor cells (PLP2 cell line). Regarding the antibacterial activity, linseed oil 3 and pumpkin oil showed the highest activity against most of the tested bacteria (especially against *Enterobacter cloacae* and *Escherichia coli*) with MIC and MBC values similar to the used positive controls (E211 and E224). All the tested oils showed significant antifungal activities, especially luffa and pumpkin oil, and for most of the tested fungi they were more effective than the positive controls, as for example in the case of *Aspergillus versicolor*, *A. niger*, and *Penicillium verrucosum* var. *cyclopium*. In conclusion, the results of our study showed promising antimicrobial and cytotoxic properties for the studied seed oils which could be partly attributed to their fatty acids composition, especially the long-chain ones with 12–18 carbons.

Keywords: seed oils; antibacterial properties; cytotoxicity; antifungal properties; omega-3 fatty acids; omega-6 fatty acids; antitumor activities; *Portulaca oleracea* L.; *Luffa aegyptica* Mill.; *Cucurbita maxima* L.; *Linum usitatissimum* L.

1. Introduction

The food industry is looking for novel natural compounds to be used as non-synthetic antimicrobial agents in the whole food chain, aiming to fulfill consumers demands for healthy and chemical-free food products [1]. The concept of bio-preservation through the use of plant derived antimicrobial agents is of major importance for food safety and food security, while providing additional functional properties to food products at the same time [2]. For this reason, the research interest has focused on various plant species also considering obtaining novel compounds from the by-products of food processing industries [3]. In this context, lipids and oils are a promising source of antimicrobial agents since according to the literature they have been found to possess such activities [4,5]. Especially for seed oils and extracts, there are several examples of significant in vitro antimicrobial effects against food-borne pathogens for seed oils of conventional and unconventional crops that could found in applications in the food industry [2,6].

Vegetable oils are considered a basic ingredient in many dietary patterns, such as the Mediterranean diet, and provide essential nutrients and valuable bioactive compounds with beneficial health effects [7]. They are considered rich sources in sterols, polyunsaturated fatty acids (PUFA), polyphenols, and carotenoids, although there could be significant differences in oil composition depending on the extraction method e.g., cold or thermal pressing, use of solvents, supercritical extraction, etc. [8]. The most common oils around the world are obtained from palm, soybean, rapeseed, and sunflower, being used for culinary purposes in raw form or after processing for the production of hydrogenated oil with further uses in the food industry, as well as in industrial applications in cosmeceuticals, paint industry, or biofuel production [9].

Apart from the well-known and widely used vegetable oils, there are several minor seed oils which possess special features related to their chemical composition and bioactive properties [10,11]. Their unique characteristics are usually related with fatty acids composition, especially omega-3 fatty acids content [12,13], or specific bioactive compounds such as polyphenols or tocopherols that contribute to the observed antioxidant and bioactive properties [14,15]. However, despite their beneficial effects, most of these oils are underexploited since either they used for industrial applications with low added value or they considered as byproducts because plants are cultivated for vegetative tissues and harvested before entering the reproductive stage e.g., fibers in the case of flaxseed [16]. During the last years, scientific research has indicated various species as rich sources of beneficial fatty acids, some of which are currently being used in the food and pharmaceutical industry [17]. For example, vegetable oils from species such as sacha inchi (*Plukenetia volubilis*), linseed (*Linum usitatissimum*), and perilla (*Perilla frutescens*) are rich in α-linolenic acid and monounsaturated fatty acids (MUFA), whereas chia (*Salvia hispanica*) and purslane (*Portulaca oleracea*) contain α-linolenic and linoleic acid (LA) in amounts that account to omega-6/omega-3 ratios with values lower than 4, indicating high nutritional value [17–19].

Flax or linseed (*Linum usitatissimum* L.) can be cultivated for its seeds and in this case it is named after the terms "linseed", "flax seed", "oil flax", or "seed flax", or it can be cultivated for its fibers and then it is referred to "flax". There are also dual purpose cultivars which can be grown both for fibers and seeds [20]. The seed oil can be used for non-edible industrial applications as a drying oil or for edible purposes due to its high content in omega-3 α-linolenic acid [20]. Due to susceptibility to lipid oxidation, it has been suggested that edible linseed oil could be encapsulated or supplemented with powders of edible flowers that increased oxidative stability and improved the quality parameters of oil [20]. Moreover, encapsulated oils broaden the applications in the food industry since they can be incorporated in various food products without altering their physicochemical properties while at the same time they improve their functional properties [21]. The high content of linseed oil in omega-3 fatty acids is also responsible for significant antioxidant, antimicrobial properties and various health effects [22], and according to Wrona et al. [23] linseed oil could be used in active packaging of food products and extend their shelf-life.

Pumpkin or cucurbit seeds are highly appreciated for their health effects and increasing consumption trends are evidenced during the last few years [24]. Seeds are usually discarded during fruit processing, therefore finding new alternative uses may increase the added value of pumpkin crop in the circular economy context. However, there is a limited cultivation of pumpkins intended for seed production and applications in the food industry as confectionaries or consumed in raw of roasted form [25]. Moreover, seeds can be used for oil extraction which exhibits several beneficial health effects due to their content in bioactive compounds such as fatty acids, tocopherols, and phytosterols (mostly squalene) [26,27]. However, extraction protocols and processing condition, as well as the genotype, may affect bioactive compounds content, and cold-pressing oils are considered of higher quality and nutritional value than thermal- or solvent-processed ones, despite the lower oil yields [28–30]. Moreover, among the various health effects seed oils also possess antimicrobial properties which may find practical application in the food and pharmaceutical industries, since they exhibit significant potency against various microbes [31,32].

On the other hand, purslane is mostly appreciated for its edible leaves and stems which are among the richest plant sources of α-linolenic acid [33]. Several health benefits have been attributed to purslane plant parts, including antioxidant, anticancer, hepatoprotective, and antimicrobial properties, among others [34,35]. However, the aerial plant parts may also contain high amounts of oxalic acid which is an antinutritional factor and may have severe effects in individuals with kidney problems [36]. Seeds and seed oils are not so commonly used and recent research highlighted their valuable properties associated with the unique chemical composition, especially the fatty acids profile [8,10], while they do not contain oxalic acid which enables their safe consumption [11,37]. Among the studied species luffa is the least known and is commonly used for its spongy flesh as a natural sponge, while other suggested uses include oil removal from waste water and oil spills [38] and biofuel production [39]. Moreover, luffa seeds are considered a rich source of protein while seeds and seed oil are edible and possess a high nutritional value and bioactive properties [40–42].

Considering the importance of vegetable oils in the human diet and the lack of information regarding less conventional sources of oil, the aim of the present study was to evaluate the fatty acids composition and the antimicrobial and cytotoxic properties of vegetable oils obtained from purslane, pumpkin, luffa, and linseed seeds.

2. Materials and Methods

2.1. Oil Samples

Oil samples were obtained from seeds of purslane (*Portulaca olearacea* L.), cucurbit (*Cucurbita maxima* L. cv. Nychaki), luffa (*Luffa aegyptica* Mill.) and linseed (*Linum usitatissimum* L.). In particular, purslane and pumpkin oils were obtained from seeds harvested from plants cultivated at the experimental farm of the University of the Thessaly in the growing period of spring–autumn 2020. Purslane plants were cultivated following the agronomic practices previously described by Petropoulos et al. [11]. In brief, seeds were sown directly in soil on June 1 2020 and harvested approximately two months later and when most of the plants reached fully maturity (3 August 2020). A base dressing with 100 kg/ha of N-P-K was applied, while plants were irrigated via a sprinkler irrigation system at regular intervals. Plant distances were 30 cm × 5 cm, between and within the rows, respectively, while the dimensions of the experimental plots were 3 m × 3 m (n = 3). No pesticides, fungicides, or herbicides were applied during the growing period, since no certified agrochemicals are available for the species. Weed control was implemented with hand hoeing. Plants were cut at 5–10 cm above ground with a scythe and seeds were removed from fruit after shaking and sieving. After harvesting, seeds were stored at dry conditions until oil extraction.

Similarly, pumpkin seeds were sown directly in soil in single rows on 27 July 2020, and fruit was harvested on 7 December 2020. Plant distances were 1.5 m within each row and 2.5 m between the rows, while three rows (n = 3) of 50 m long were used (33 plants per row and 100 plants in total). A base dressing with 250 kg per hectare of 12-11-18 (N-P-K) fertilizer (YaraMila Complex, Yara Hellas S.A., Greece) was applied before sowing, while during the growing period plants were fertigated with ammonium nitrate (34.5% nitrogen at 40 kg per hectare; Idealfer, Gavriel S.A., Greece) and calcium nitrate (15.5-0-0 +26% CaO at 50 kg per hectare; Haifa-Cal, Haifa Group, Greece). Plants were irrigated via drip irrigation system at regular intervals. Pests and diseases management was carried according to best practice guides, while for weed control mulching with plastic film on sowing lines was applied. After the harvest, fruit were dissected and seeds were removed from the pulp, washed with distilled water, and left to dry at room temperature. After drying, seeds were stored at dry conditions until oil extraction.

Linseed and luffa seed oils were obtained from local retail shops and from Giachanas—Cold Pressed Seed Oils S.A. (Evros, Greece). In particular, three different samples of linseed and one sample of luffa seed oil were studied, namely linseed oil 1 (Chemco Linseed Oil; Syndesmos S.A., Greece), linseed oil 2 (Giachanas—Cold Pressed Seed Oils S.A.), linseed oil 3 (Organic flaxseed oil; Biotuscany srl., Italy), and luffa seed oil (Giachanas—Cold Pressed Seed Oils S.A.). All commercially available seed oils were extracted via cold pressing, as indicated on the product label.

For purslane and pumpkin seed oils (Purslane oil and Pumpkin oil, respectively), extraction was carried out via cold pressing according to the methodology and the conditions previously described by the authors [10,11]. Both oils were extracted by Giachanas—Cold Pressed Seed Oils S.A., while the cold press was a Henan 6 YL-160 screw oil press (Henan VIC Machinery Co. Ltd.; Henan, China) [11].

After extraction, oils were stored at room temperature at dry and dark conditions in amber screw top glass vials until further analyses.

2.2. Chromatographic Analysis of Fatty Acids

Fatty acids were determined by gas–liquid chromatography with flame ionization detection (GC-FID), after the crude fat extraction and derivatization procedures described by Spréa et al. [43]. The analysis was carried out with a DANI model GC 1000 instrument equipped with a split/splitless injector, a FID, and a Macherey-Nagel column. Fatty acid identification was made by comparing the relative retention times of fatty acid methyl ester (FAME) peaks from samples with standards. The results were recorded and processed using the Clarity DataApex 4.0 Software and expressed in relative percentage of each fatty acid.

2.3. Evaluation of Bioactive Properties in vitro

2.3.1. Preparation of Oil Extracts

The oils (5 mL) were extracted by liquid–liquid with 10 mL of methanol, this procedure was repeated 3 times. Then, the combined extracts were dried over anhydrous sodium sulphate, and after filtration were evaporated under reduced pressure until dryness.

2.3.2. Cytotoxic Activity

The cytotoxic capacity of the extracts prepared above was assessed by the sulforhodamine B (Sigma-Aldrich, St. Louis, MO, USA) assay against a non-tumor cell line (PLP2, porcine liver primary cells) (acquired from Leibniz-Institut DSMZ). The same assay was also used to evaluate the extracts cytotoxicity against four human tumor cell lines, namely HeLa (cervical carcinoma), HepG2 (hepatocellular carcinoma), MCF-7 (breast adenocarcinoma), and NCI-H460 (non-small cell lung cancer), accordingly a procedure described by the authors [43,44]. Ellipticine (Sigma- Aldrich, St. Louis, MO, USA) was used as a positive control. The results were expressed in GI50 values (μg/mL), corresponding to the extract concentration responsible for 50% inhibition of growth in a primary culture of liver cells-PLP2 or in human tumor cell lines.

2.3.3. Antimicrobial Activity Evaluation

The following Gram (+) bacteria: *Staphylococcus aureus* (ATCC 11632), *Bacillus cereus* (food isolate), and *Micrococcus flavus* (ATCC 10240), as well as Gram (−) bacteria *Enterobacter cloacae* (ATCC 35030), *Salmonella typhimurium* (ATCC 13311), and *Escherichia coli* (ATCC 25922) were selected to test the antibacterial activity of the extracts. Fungi *Aspergillus fumigatus* (ATCC 9197), *Aspergillus versicolor* (ATCC 11730), *Aspergillus niger* (ATCC 6275), *Penicillium funiculosum* (ATCC 36839), *Penicillium verrucosum* var. *cyclopium* (food isolate), and *Trichoderma viride* (IAM 5061) were selected to test the antifungal activity of the extracts. The microorganisms were obtained from the Mycological Laboratory, Department of Plant Physiology, Institute for Biological Research "Siniša Stanković", University of Belgrade, National Institute of Republic of Serbia.

The extracts were redissolved in 5% dimethyl sulfoxide (DMSO) and the microdilution method was used as previously mentioned by the authors [45]. The results were presented as minimum inhibitory/bactericidal concentrations (MICs/MBCs) in case of bacteria and minimum inhibitory/fungicidal concentrations (MICs/MFCs) for fungi; E211 (sodium benzoate) and E224 (potassium metabisulphite) were used as positive controls and 5% DMSO was used as a negative control.

2.4. *Statistical Analysis*

Throughout this work, results are expressed as mean ± standard deviation. An analysis of variance (ANOVA) was used to determine the statistical difference among the samples. Mean and standard deviations were determined from the obtained data using Microsoft Excel (Microsoft Corp., Redmond, WA, USA). ANOVA was performed with the use of Statgraphics 5.1.plus (Statpoint Technologies, Inc., Warrenton, VA, USA).

3. Results and Discussion

The fatty acids composition of the studied seed oils is presented in Table 1, where fifteen individual fatty acids were detected. All the oils contained almost the same fatty acids, except for the case of pentadecanoic acid (C15:0) which was detected only in purslane seed oils, docosadienoic acid (C22:2) which was identified in linseed and luffa oils, and lignoceric acid (C24:0) which was found only in pumpkin oil. Polyunsaturated fatty acids were the most abundant class in all the studied oils (55.92–85.51%), followed by monounsaturated and saturated fatty acids (MUFA and SFA, respectively) in amounts that differed depending on the studied oil. This result is within the same range with previous reports regarding linseed [17], luffa [26], purslane [11], and pumpkin [17] seed oils where unsaturated fatty acids were the predominant class, although at slightly different amounts. These differences could be attributed to different extraction methods between the studies or the different genotypes tested, as already pointed out Petropoulos et al. [10,11].

Moreover, fatty acids profile varied among the studied oils. In particular, α-linolenic acid (C18:3n3) was the most abundant fatty acid in the case of linseed oils (71.90%, 70.93%, and 65.62% for linseed oil 1, 2, and 3, respectively), while the second most abundant compound was linoleic acid (C18:2n6c) followed by oleic (C18:1n9c), palmitic (C16:0), and stearic (C18:0) acids. Similarly, to our study, Zamani Ghaleshahi et al. [46] suggested α-linolenic as the most abundant fatty acid, while the next most important fatty acids were oleic, palmitic, and stearic acids. However, despite the similarities, the detected amounts in the tested samples of our study were higher than these reports, especially in the case of α-linolenic acid, indicating differences in the extraction method and probably the genotypes tested. Moreover, Tavarini et al. [20], who tested two linseed varieties (Bethune and Solal), also observed significant differences in fatty acids profile, especially in the case of α-linolenic acid which was considerably higher in Bethune compared to Solal genotype which was a low linolenic mutant (64.02 g/100 g fatty acids and 3.96 g/100 g fatty acids, respectively). These effects could be also observed in our study, where the fatty acids profile differed among the three linseed tested oils. However, according to Ren et al. [47], seed pretreatment may significantly increase extraction yield without affecting the fatty acids profile, while Gutte et al. [48] did not detect differences in fatty acids profile of oils extracted with solvent or ultrasonic assisted.

Table 1. Fatty acids composition (%) of the studied vegetable oils (mean ± SD).

Fatty Acids	Linseed Oil 1	Linseed Oil 2	Linseed Oil 3	Luffa Oil	Purslane Oil	Pumpkin Oil
C14:0	0.042 ± 0.001e	0.043 ± 0.001e	0.054 ± 0.001c	0.095 ± 0.002b	0.049 ± 0.001c	0.155 ± 0.003a
C15:0	-	-	-	-	0.028 ± 0.001	-
C16:0	4.61 ± 0.04e	4.32 ± 0.04f	5.36 ± 0.01d	13.77 ± 0.07c	14.1 ± 0.1b	14.72 ± 0.05a
C16:1	0.068 ± 0.004c	0.044 ± 0.001d	0.086 ± 0.004b	0.153 ± 0.005a	0.086 ± 0.001b	0.15 ± 0.01a
C17:0	0.057 ± 0.004e	0.055 ± 0.004e	0.072 ± 0.002c	0.166 ± 0.004a	0.105 ± 0.003c	0.113 ± 0.002b
C18:0	2.85 ± 0.01d	3.10 ± 0.01c	3.06 ± 0.01c	6.99 ± 0.02a	3.046 ± 0.006c	6.24 ± 0.01b
C18:1n9c	7.09 ± 0.01c	6.43 ± 0.01e	6.95 ± 0.01d	15.14 ± 0.03b	5.24 ± 0.02f	21.76 ± 0.01a
C18:2n6c	12.87 ± 0.01f	14.58 ± 0.01e	17.98 ± 0.02d	61.86 ± 0.01a	34.10 ± 0.05c	55.25 ± 0.05b
C18:3n3	71.90 ± 0.05a	70.93 ± 0.01b	65.62 ± 0.04c	0.94 ± 0.02f	41.25 ± 0.08e	0.323 ± 0.006g
C20:0	0.095 ± 0.002e	0.102 ± 0.001e	0.153 ± 0.001d	0.312 ± 0.003c	0.60 ± 0.01a	0.401 ± 0.005b
C20:1	0.080 ± 0.002c	0.081 ± 0.001c	0.124 ± 0.002a	0.057 ± 0.001d	0.102 ± 0.001b	0.119 ± 0.008a
C22:0	0.081 ± 0.001e	0.088 ± 0.005d	0.133 ± 0.004c	0.092 ± 0.006d	0.25 ± 0.01a	0.144 ± 0.002b
C22:2	-	-	-	-	0.49 ± 0.05a	0.35 ± 0.03b
C23:0	0.186 ± 0.004e	0.164 ± 0.002f	0.222 ± 0.005d	0.251 ± 0.001c	0.51 ± 0.01a	0.292 ± 0.003b
C24:0	0.085 ± 0.005b	0.061 ± 0.001c	0.183 ± 0.002a	0.180 ± 0.001a	-	-
SFA	8.01 ± 0.04e	7.94 ± 0.02f	9.24 ± 0.01d	21.86 ± 0.06b	18.7 ± 0.1c	22.06 ± 0.06a
MUFA	7.24 ± 0.01c	6.56 ± 0.01d	7.16 ± 0.01d	15.35 ± 0.03b	5.43 ± 0.02e	22.03 ± 0.02a
PUFA	84.76 ± 0.04b	85.51 ± 0.01a	83.60 ± 0.02c	62.79 ± 0.04e	75.83 ± 0.08d	55.92 ± 0.08f
PUFA/SFA	10.58 ± 0.02b	10.77 ± 0.01a	9.05 ± 0.01c	2.87 ± 0.05e	4.06 ± 0.04d	2.55 ± 0.07f
n6/n3	0.18 ± 0.03e	0.21 ± 0.01e	0.27 ± 0.03d	65.81 ± 0.02b	0.83 ± 0.06c	166.57 ± 0.03a

C14:0 myristic acid; C15:0 pentadecanoic acid; C16:0 palmitic acid; C16:1 palmitoleic acid; C17:0 heptadecanoic acid; C18:0 stearic acid; C18:1n9c oleic acid; C18:2n6c linoleic acid; C18:3n3 α-linolenic acid; C20:0 arachidic acid; C20:1 eicosenoic acid; C22:0 behenic acid; C22:2 docosadienoic acid; C23:0 tricosylic acid; C24:0 lignoceric acid; SFA: saturated fatty acids; MUFA: monounsaturated fatty acids; PUFA: polyunsaturated fatty acids; n6/n3: ratio of omega-6/omega-3 fatty acids; -: not detected. In each row, different letters mean statistical differences among samples.

In the case of luffa seed oil, it included mostly linoleic acid (61.86%) followed by oleic acid and palmitic detected in similar amounts (15.14% and 13.77%, respectively), and stearic acid which was found in lesser amounts (6.99%). A similar profile was reported by Stevenson et al. [26] and Adewuyi et al. [39], although they reported a lower content of linoleic acid compared to our study (43.7%, 46.8%, and 46.1%, respectively). It seems that as in the case of linseed oil the detected differences could be attributed to the extraction methods since Stevenson et al. [26] extracted oil via supercritical and solvent extraction, while Adewuyi et al. [39] implemented the Soxhlet extraction method.

Purslane seed oil was also a rich source of α-linolenic acid (41.25%) and linoleic acid (34.10%), followed by palmitic, oleic, and stearic acids (14.1%, 5.2%, and 3.0%, respectively). In contrast, Dubois et al. [17] reported a slightly higher content of linoleic than α-linolenic acid (34.1% and 32.4%, respectively), while Liu et al. [49] suggested significant differences between wild and cultivated genotypes of purslane, with wild ones containing significantly higher amounts of α-linolenic than linoleic acid. Regarding the extracting method, Sodeifian et al. [8] did not observe any differences in fatty acids profile between the Soxhlet and solvent extraction methods, whereas Petropoulos et al. [11] suggested the extraction conditions (e.g., cold extraction vs heat extraction) may affect the content of α-linolenic acid and consequently improve the nutritional value and bioactive properties of oil. However, this was not the case in the study of Delfan-Hosseini et al. [50] who compared solvent extraction with cold-pressing and the combination of seed pretreatment with microwaves prior to cold-pressing, and did not observe any differences. This contradiction could be attributed to the fact that cold-pressing in the later report refers to room conditions where extraction took place, since no cooling module was described as in the case of Petropoulos et al. [11]. According to Ionescu et al. [51], several operational parameters (e.g., screw speed, press nozzle size, and pressure), may affect not only extraction yield but also oil composition, since they affect temperature conditions during extraction.

Finally, pumpkin oil showed similarities with luffa oil and contained slightly less linoleic and more oleic acid (55.25% and 21.76%, respectively), while palmitic and oleic acids were found in amounts similar to luffa oil. The detected fatty acid profile in pumpkin seed oil of our study was similar to the one suggested by Nederal et al. [29], whereas Nyam et al. [52] reported higher amounts of oleic than linoleic acid in *Cucurbita pepo* seeds (42.8% and 30.4%, respectively). Considering the numerous genotypes and the differences in fatty acid profiles among the seed oils of various cultivars, it is difficult to suggest a specific composition for pumpkin oil, since there are genotypes with similar amounts of linoleic and oleic acid, while others contain more linoleic than oleic acid, as in the case of our study [26]. Moreover, according to Murkovic et al. [53], linoleic acid content are highly correlated to each other and linoleic acid is formed after the dehydrogenation of oleic acid. In the same study, it was also suggested that earliness in maturity of pumpkin fruit and environmental conditions may also affect fatty acid composition in seeds and consequently in seed oils [53]. Therefore, late harvesting under low temperatures may result in higher linoleic acid and polyunsaturated fatty acids content due to higher activity of oleoyl phosphatidylcholine desaturase [54].

The values of PUFA/SFA ratio were higher than 0.45 for all the studied oils, indicating a high nutritional value, especially in the case of linseed oils where the highest values were recorded (9.05–10.77), whereas pumpkin oil values were marginally higher than this threshold [55–57]. Similarly, n6/n3 ratio was lower than 4.0 in linseed and purslane seed oils due to their high content in omega-3 fatty acids and α-linolenic acid in particular. On the other hand, the values of n6/n3 recorded for luffa and pumpkin seed oils were considerably higher than 4.0 due to their high content of omega-6 fatty acids (linoleic acid) and the very low amounts of α-linolenic acid. The recorded values for both ratios are in agreement with the literature reports as already described [17,26,29,46,53], except for the cases where genotypic differences or differences in the extraction protocols were identified [20,49,52]. According to Guil et al. [55], both these ratios (PUFA/SFA and n6/n3) are good indicators for the nutritional quality of a food product, however even in the case of pumpkin and luffa oils which did not met this specific criterion, there are several beneficial health effects evidenced that allow us to suggest their incorporation in the human diet. Moreover, it seems that the extraction with cold pressing may improve the nutritional value of seed oils by increasing the beneficial fatty acids content, such as α-linolenic acid in the case of linseed and purslane seed oil, thus improving the nutritional value and the bioactive properties of the obtained oils [58].

The cytotoxic effects of the studied oils are described in Table 2. None of the tested oils were effective against all the tested tumor cell lines, while linseed oil 1, luffa oil, and cucurbit showed a slight toxicity against the non-tumor porcine liver primary culture (PLP2) cell line. Moreover, all the tested oils (except for linseed oil 1 and 2) showed efficacy against cervical carcinoma (HeLa) cell lines, especially luffa oil which recorded the lowest GI_{50} values (215 μg/mL), followed by pumpkin oil, linseed oil 1, and purslane oil in decreasing order of effectiveness. Similarly, luffa oil was the most efficient against non-small cell lung cancer (NCI-H460) cell line, followed by linseed oil 1, whereas none of the tested oils were effective against hepatocellular carcinoma (HepG2) and breast carcinoma (MCF-7) cell lines. The anticancer activities of *Luffa* sp. aerial parts against various types of cancer have been previously reported (e.g., human neuronal glioblastoma cells (U343) and human lung cancer cells (A549) [59], human lung cancer cell line (NCI-H460) [60], Ehrlich ascites carcinoma (EAC) [61], Dalton's lymphoma ascites (DLA) [62]); however, to the best of our knowledge there are no reports regarding the cytotoxic effects of luffa seed oil against cancer cell lines and the results of our study could be useful for the exploitation of this underutilized species. Linseed is also well known for its anticancer activities, especially against hormone related cancers such as breast, prostate, and colon cancer, due to its high content in α-linolenic acid and polyphenols which reduce human estrogen receptor-positive tumors and inhibit cancer cells proliferation [63–65]. Therefore, our results are in agreement with those of literature reports, although it seems that bioactive properties of oils are related to the extraction protocol since not all the tested oils of this study presented similar cytotoxic effects.

Table 2. Cytotoxicity and antitumor activity of the studied vegetable oils (GI_{50} values μg/mL).

Seed Oil	Cytotoxicity to Non-Tumor Cell Lines	Cytotoxicity to Tumor Cell Lines			
	PLP2 (Porcine Liver Primary Culture)	**HeLa (Cervical Carcinoma)**	**HepG2 (Hepatocellular Carcinoma)**	**MCF-7 (Breast Carcinoma)**	**NCI-H460 (Non-Small Cell Lung Cancer)**
Linseed oil 1	301 ± 23a	291±27b	>400	>400	369 ± 33a
Linseed oil 2	>400	>400	>400	>400	>400
Linseed oil 3	>400	>400	>400	>400	>400
Luffa oil	215 ± 17c	189 ± 17c	>400	>400	136 ± 12b
Purslane oil	>400	307 ± 12a	>400	>400	>400
Pumpkin oil	259 ± 21b	270 ± 25b	>400	>400	>400

GI_{50} values correspond to the sample concentration responsible for 50% inhibition of growth in a primary culture of liver cells-PLP2 or in human tumor cell lines or. GI_{50} values for Ellipticine (positive control): 3 μg/mL (PLP2), 1.0 μg/mL (MCF-7), 1.0 μg/mL (NCI-H460), 2.0 μg/mL (HeLa) and 1.0 μg/mL (HepG2). In each row, different letters mean statistical differences among samples.

Regarding pumpkin seeds and seed oil, both in vitro and in vivo studies have reported the beneficial effects against hyperplastic and cancer cells, as in the case of prostate hyperplasia [66], colon cancer [67], or cervical carcinoma [68]. According to Colagar and Souraki [69] the anticancer effects of pumpkin seed oil could be attributed to antioxidant vitamins which inhibit free radicals production, while Heng et al. [70] associated toxic effects against melanoma cells to moschatin, a ribosome inactivating protein. In a recent study conducted by Khan et al. [71], cucurbitacin obtained from seeds exhibited antiproliferative effects against non-small cell lung cancer cells, while Bardaa et al. [72] reported that bioactive properties of seed oil could be due to high content not only of polyunsaturated fatty acids, but also to tocopherols and phytosterols. Moreover, Al-Okbi et al. [73] reported significant inhibitory effects of *Cucurbita moschata* and *C. pepo* seed oils against liver, colon, and breast cancer, and suggested potential uses in controlling tumor proliferation.

The lack of toxic effects or genotoxicity of purslane has been previously reported for aqueous extracts of aerial plant parts, indicating they are safe for edible purposes [74], while moderate toxic effects were suggested for solvent extracts against human lung (K562 and A549) and breast (MCF-7 and MDA-MB-435) cancer cell lines [75], nasopharyngeal cancer (CNE-1), colon cancer (HT-29), and cervical cancer (HeLa). The abovementioned reports refer to aerial plant parts (stems and leaves), while the work of Al-Sheddi et al. [76] suggested significant cytotoxic and anti-proliferative effects of purslane seed extracts or seed oil against HeLa and A549 cell lines. According to Al-Sheddi et al. [76] purslane seed oil may exert significant in vitro cytotoxic effects against cell lines of human liver (HepG2) and human lung cancer (A-549). Moreover, Asif suggested that the anticancer activities of seed oils could be attributed to polyunsaturated fatty acids, especially a group of conjugated isomers of linoleic acid [5].

The antibacterial properties of the tested oils are presented in Table 3. A varied effectiveness was recorded against the tested bacteria depending on the oil source. In particular, all the oils exhibited high bactericidal and growth inhibitory effects against *S. aureus* and *M. flavus* with MIC and MBC similar to E211 and E224 (positive controls), respectively (except for linseed oil 2 and purslane oil, which showed the lowest activity in the case of *S. aureus* and *M. flavus*, respectively). Moreover, all the tested oils were similarly or more effective than E224 against *B. cereus*, while E211 was the most effective. Linseed oil 2 and luffa oil recorded MIC values similar to E224 against *E. cloacae*, while (with the exception of purslane oil) the rest of the tested oils were more effective than the other positive control (E211). Regarding *S. typhimurium*, the tested oils (except for linseed oil 2 and purslane oil) were similarly effective to positive controls, apart from E224 which recorded the lowest MBC values. Finally, linseed oil 3 and pumpkin oil showed higher effectiveness against *E. coli* than the rest of the oils and similar to E224.

The antibacterial activities of flaxseed oil are well documented and variable effects have been suggested against bacteria, such as methicillin resistant *S. aureus* (MRSA), methicillin sensitive *S. aureus* (MSSA), *K. pneumoniae* and *S. epidermidis*, whereas no effectiveness against *E. coli* and *E. faecalis* was observed [77]. Similarly, Kaithwas et al. [78] suggested effectiveness of flaxseed oil against a broad spectrum of bacteria (e.g., *S. aureus*, *Streptococcus agalactiae*, *E. coli*, *E. faecalis*, and *M. luteus*) was with antimicrobial activity comparable to or better than the positive control (cefoperazone). According to Santos et al. [79], flaxseed oil exhibited significant antibacterial activity against *Salmonella enteritidis* and *S. typhimurium* and no inhibitory effects against *S. aureus*, *L. monocytogenes*, and *B. cereus*. The same authors also suggested that cultivation practices may affect the bioactive properties of the obtained oils, with oils extracted from seeds of organically grown plants being more potent than those of conventionally grown ones [79]. This seems to be the case in our study, since linseed oil 3 (organic oil) was more effective against *E. coli* than the other two linseed oils, while for the rest of the tested bacteria no differences were observed in MIC and MBC from linseed oil 2. Moreover, Joshi et al. [80] suggested the complementary use of flaxseed oil along with gemifloxacin to increase its effectiveness and reduce the development of resistance.

Table 3. Antibacterial activity of the studied seed oils (MIC and MBC mg/mL).

Seed Oil	*Staphylococcus aureus* (ATCC 11632)		*Bacillus cereus* (Food Isolate)		*Micrococcus flavus* (ATCC 10240)		*Enterobacter cloacae* (ATCC 35030)		*Salmonella* Typhimurium (ATCC 13311)		*Escherichia coli* (ATCC 25922)	
	MIC	MBC	MIC	MBC	MIC	MBC	MIC	MBC	MIC	MBC	MIC	MBC
Linseed oil 1	2.00	4.00	1.00	2.00	1.00	2.00	1.00	2.00	1.00	2.00	1.00	2.00
Linseed oil 2	4.00	8.00	1.00	2.00	1.00	2.00	0.50	1.00	2.00	4.00	2.00	4.00
Linseed oil 3	2.00	4.00	1.00	2.00	1.00	2.00	1.00	2.00	1.00	2.00	0.50	1.00
Luffa oil	2.00	4.00	1.00	2.00	1.00	2.00	0.50	1.00	1.00	2.00	2.00	4.00
Purslane oil	2.00	4.00	2.00	4.00	2.00	4.00	2.00	4.00	2.00	4.00	1.00	2.00
Cucurbit oil	2.00	4.00	1.00	2.00	1.00	2.00	1.00	2.00	1.00	2.00	0.50	1.00
E211	4.00	4.00	0.50	0.50	1.00	2.00	2.00	4.00	1.00	2.00	1.00	2.00
E224	1.00	1.00	2.00	4.00	1.00	2.00	0.50	0.50	1.00	1.00	0.50	1.00

MIC: minimum inhibition concentration; MBC: minimum bactericidal concentration; E211: sodium benzoate; E224: potassium metabisulphite.

The studies of antibacterial effects of luffa refer to aerial plant parts and seeds with limited research available on seed oil effects [42,62]. According to Swain et al. [81], luffa seed oils showed effectiveness against a wide range of bacteria (e.g., *S. aureus*, *S. epidermidis*, *Micrococcus leutius*, *Pseudomonas aeruginosa*, and *E. coli*), while in all the cases the obtained MIC values were lower than the positive control used. Similarly, the reports for the antibacterial effects of purslane are limited, especially for seed oils [82–85]. According to Tayel et al. [86], purslane seed extracts were only effective against normal *S. aureus* strains, while Bakkiyaraj and Pandiyaraj [87] also recorded effectiveness of leaf extracts against *S. aureus*, *B. aureus*, and *P. aeruginosa*. Recently, Petropoulos et al. [11] reported low effectiveness of purslane seed oil against different Gram+ and Gram- bacteria, although they suggested that raw seeds were more effective than seedcakes and seed oils, probably due to loss of bioactive compounds during the extraction procedure. In contrast, Othman [88] found significant effectiveness of purslane fixed oil against *S. epidermidis* and *E. coli* and attributed this activity to the high content of omega-3 fatty acids.

Regarding pumpkin oil, Obi et al. [32] reported significant inhibitory effects of *Cucurbita pepo* seed oils against *S. aureus* and *E. coli* and attributed these effects to the presence of bioactive compounds such as tannins, flavonoids, cyanogenic glycosides, cardiac glycosides, and saponins. Moreover, Bardaa et al. [72] reported significant inhibitory effects of pumpkin seed oil against *Bacillus subtilis*, while Amin et al. [31] suggested high effectiveness against various *E. coli* strains and *Shigella sonnei* and further reported significantly higher antibacterial activity in oils obtained from an indigenous genotype compared to a hybrid. Moreover, in an earlier study of the same authors, the oil of indigenous pumpkin seeds contained higher amounts (2.5 times higher) of tocopherols than a hybrid pumpkin, a finding that could justify the better antibacterial efficacy of this particular oil [31]. In contrast, Hammer et al. [89] classified pumpkin oil among the plant extracts that did not show any inhibitory effects against several bacteria when applied at the highest tested concentration (2.0%; v/v).

From the results of this study it could be concluded that fatty acids composition is essential for the antibacterial properties of the tested oils. In their review report, Yoo et al. [58] highlighted the broad spectrum of antibacterial activity of lipids with emphasis on fatty acids with 18 carbons such as α-linolenic, linoleic, and oleic acid, which were the most prevalent fatty acids in the oils of our study as well. Moreover, Xuan et al. [90], who studied antibacterial effects of various edible oils, highlighted the differences in total phenolic and total flavonoids content as well as to individual compounds of the tested oils as responsible for the varied antibacterial efficacy. However, despite the differences in chemical composition, there was no positive correlation of phenolics and flavonoids with antibacterial effects against *S. aureus* and *E. coli*. Moreover, the processing grade may also affect antibacterial effects with crude oils being more effective than the refined ones, indicating the loss of valuable compounds during the refining process [91]. These findings suggest that other antioxidant compounds apart from lipids are responsible for the antibacterial effects of vegetable oils (e.g., tocopherols or species specific compounds) which could justify the observed differences among the studied oils of our study. Moreover, agronomic practices and the extraction method may also have an effect on bioactive properties of seed oils, as evidenced in the case of our study with different linseed oils.

The antifungal properties of the seed oils of our study are shown in Table 4 with the tested oils being more effective than the positive controls in several occasions. In particular, linseed oil 1 and 3 were more effective against *A. fumigatus* compared not only to the rest of the tested oils, but also to the used controls. Similarly, luffa and cucurbit oils were the most effective against *A. versicolor*, *A. niger*, *P. funiculosum*, and *P. verrucosum* var. *cyclopium*, without differences from linseed oil 3 in the case of *A. niger*, and linseed oil 2 in the case of *P. funiculosum* and *P. verrucosum* var. *cyclopium*. Finally, linseed oil 2 and luffa oil were the most effective against *T. viride* with MIC values equal to E224. Purslane seed oil and linseed oil 3 (except for the case of *A. niger*) had the least overall effectiveness against the tested fungi, since in most cases they recorded the highest MIC and MFC values.

Table 4. Antifungal activity of the studied seed oils (MIC and MFC mg/mL).

Seed Oil	*Aspergillus fumigatus* (ATCC 9197)		*Aspergillus versicolor* (ATCC 11730)		*Aspergillus niger* (ATCC 6275)		*Penicillium funiculosum* (ATCC 36839)		*Penicillium verrucosum* var. *cyclopium* (Food Isolate)		*Trichoderma viride* (IAM 5061)	
	MIC	MFC	MIC	MFC	MIC	MFC	MIC	MFC	MIC	MFC	MIC	MFC
Linseed oil 1	0.50	1.00	1.00	2.00	2.00	4.00	2.00	4.00	1.00	2.00	1.00	2.00
Linseed oil 2	2.00	4.00	1.00	2.00	1.00	2.00	0.5	1.00	0.5	1.00	0.5	1.00
Linseed oil 3	1.00	2.00	2.00	4.00	0.50	1.00	2.00	4.00	2.00	4.00	2.00	4.00
Luffa oil	0.50	1.00	0.50	1.00	0.50	1.00	0.50	1.00	0.50	1.00	0.50	1.00
Purslane oil	4.00	8.00	2.00	4.00	2.00	4.00	2.00	4.00	2.00	4.00	2.00	4.00
Cucurbit oil	1.00	2.00	0.50	1.00	0.50	1.00	0.50	1.00	0.50	1.00	1.00	2.00

Table 4. *Cont.*

Seed Oil	*Aspergillus fumigatus* (ATCC 9197)		*Aspergillus versicolor* (ATCC 11730)		*Aspergillus niger* (ATCC 6275)		*Penicillium funiculosum* (ATCC 36839)		*Penicillium verrucosum* var. *cyclopium* (Food Isolate)		*Trichoderma viride* (IAM 5061)	
	MIC	MFC	MIC	MFC	MIC	MFC	MIC	MFC	MIC	MFC	MIC	MFC
E211	1.00	2.00	2.00	2.00	1.00	2.00	1.00	2.00	2.00	4.00	1.00	2.00
E224	1.00	1.00	1.00	1.00	1.00	1.00	0.50	0.50	1.00	1.00	0.50	0.50

MIC: minimum inhibition concentration; MFC: minimum fungicidal concentration; E211: sodium benzoate; E224: potassium metabisulphite.

Most of the reports regarding the antimicrobial effects of vegetable seed oils refer to antibacterial activity and limited research is carried out on fungicidal effects. Similarly to our study, Abdelillah et al. [92], who tested the fungicidal activity of linseed oil, reported effectiveness against toxigenic *Aspergillus* species and further attributed these antifungal effects to the high content of linoleic and α-linolenic acids. Swain et al. [81] also recorded higher activity of luffa oil against various fungi strains (e.g., *C. albicans*, *C. tropicalis*, *Trichophyton rubrum*, *Epidermophyton flocossum*, and *Microsporum canis*) compared to fluconazole and aqueous or ethanolic extracts of fruit. Recently, Amrithaa and Geetha [93] reported in vitro inhibitory activity of pumpkin oil against *Candida albicans* when added at 200 μL, while Abd El-Aziz et al. [94] suggested effectiveness against yeast species such as *C. albicans* and *Rhodotorula rubra* and lower effectiveness against mold species (e.g., *Penicillium chrysogenum*, *A. parasiticus*, and *A. niger*). Similar results were reported by Sener et al. [95] and Kaithwas et al. [78], who also detected high efficiency of *Cucurbita pepo* seed oil against *C. albicans*. In contrast, Hammer et al. [89] did not observe any inhibitory effects of pumpkin oil against several bacteria when applied at the highest tested concentration (2.0%; *v*/*v*). According to Rabrenović et al. [24], the high content of squalene in pumpkin seed oil could be partly responsible for these antifungal effects, apart from polyunsaturated fatty acids. Similarly to our study, Petropoulos et al. [11] recorded low effectiveness of purslane seed oil against the same bacteria since positive controls exhibited lower MIC and MBC values. It seems that lipids are responsible for the antifungal effects of the studied oils vegetable oils, while other compounds (e.g., squalene in the case of pumpkin oil) may also contribute to the overall antimicrobial activity of seed oils. Moreover, similar to antibacterial, the growing conditions may also affect the antimicrobial properties of the studied oils.

The range of all the tested pathogens used in the study is wide and includes both clinically relevant pathogens and food/crop contaminants, in order not to overlook the extensive potential of natural oils in both food and pharmaceutical industries. Hence, the obtained data, which clearly demonstrates antimicrobial potential towards all the tested pathogens, point to the fact that oils are rich in compounds with this activity. Given that previously published data have shown polyunsaturated acids have significant antibacterial and antifungal potential, it is safe to say that the antimicrobial potential observed for the oils in this study, in part, can be attributed to them.

4. Conclusions

Based on the results of our study, the tested less-conventional seed oils showed promising nutritional value regarding their fatty acids profile, with linseed and purslane seed oils having high amounts of health beneficial α-linolenic acid. On the other hand, pumpkin and luffa oil were the most abundant in linoleic acid, which is also associated with beneficial health effects. This could be supported by the in vitro cytotoxic activity of luffa oil against cervical carcinoma and non-small cell lung cancer cell lines. The tested oils also showed a varied effectiveness against several Gram+ and Gram- bacteria, especially linseed oil 3 and pumpkin oil, which showed the significant activity against most of the tested bacteria (especially against *Enterobacter cloacae* and *Escherichia coli*), although in most of the cases the positive controls exhibited the lowest MIC and MBC values. On the other hand, the antifungal activities were more profound, and the studied oils were more effective than the positive controls for most of the tested fungi, especially luffa and pumpkin oil, which showed the best overall performance. Therefore, the studied oils could be used as a source of compounds with antimicrobial potential but have also been exploited for their high nutritional value and cytotoxic effects as a functional ingredient in food products, thus increasing the added value of the corresponding crops. Moreover, further research is needed in order to evaluate the physicochemical properties and bioactive compounds in blends of the studied oils or in blends with other conventional vegetable oils, since synergistic effects may improve the overall bioactive properties of conventional oils.

Author Contributions: Conceptualization, S.A.P., Y.R., I.C.F.R.F., and L.B.; methodology, S.A.P. Â.F., R.C.C., J.P.; software, Â.F., R.C.C., J.P.; validation, Â.F., R.C.C., J.P.; formal analysis, Â.F., R.C.C., J.P.; investigation, Â.F., R.C.C., J.P.; resources, S.A.P., M.S., I.C.F.R.F., and L.B.; data curation, S.A.P., Â.F., R.C.C., J.P.; writing—original draft preparation, S.A.P., Y.R.; writing—review and editing, S.A.P., Y.R., M.S., I.C.F.R.F., and L.B.; visualization, S.A.P.; supervision, S.A.P., M.S. and L.B.; project administration, S.A.P., M.S., L.B.; funding acquisition, S.A.P., Â.F., M.S., L.B. All authors have read and agreed to the published version of the manuscript.

Funding: This work was funded by the General Secretariat for Research and Technology of Greece and PRIMA foundation under the project PULPING (Prima2019-08). The authors are grateful to the Foundation for Science and Technology (FCT, Portugal) for financial support through national funds FCT/MCTES to CIMO (UIDB/00690/2020); for the financial support through national funding from the FCT, within the scope of the Project PRIMA Section 2—Multi-topic 2019: PulpIng (PRIMA/0007/2019); and L. Barros and Â. Fernandes thank the national funding by FCT, P.I., through the institutional scientific employment program-contract for their contracts. The authors are also grateful to the Ministry of Education, Science and Technological Development of the Republic of Serbia, grant number 451-03-9/2021-14/ 200007.

Institutional Review Board Statement: Not applicable.

Informed Consent Statement: Not applicable.

Data Availability Statement: The study did not report any data.

Conflicts of Interest: The authors declare no conflict of interest.

References

1. Saeed, F.; Afzaal, M.; Tufail, T.; Ahmad, A. Use of Natural Antimicrobial Agents: A Safe Preservation Approach. In *Active Antimicrobial Food Packaging*; Var, I., Uzunlu, S., Eds.; IntechOpen: London, UK, 2019; Volume 32, pp. 1–18.
2. Quinto, E.J.; Caro, I.; Villalobos-Delgado, L.H.; Mateo, J.; De-Mateo-silleras, B.; Redondo-Del-río, M.P. Food safety through natural antimicrobials. *Antibiotics* **2019**, *8*, 208. [CrossRef]
3. Herrero, M.; Cifuentes, A.; Ibañez, E. Sub- and supercritical fluid extraction of functional ingredients from different natural sources: Plants, food-by-products, algae and microalgae—A review. *Food Chem.* **2006**, *98*, 136–148. [CrossRef]
4. Isaacs, C.E.; Litov, R.E.; Thormar, H. Antimicrobial activity of lipids added to human milk, infant formula, and bovine milk. *J. Nutr. Biochem.* **1995**, *6*, 362–366. [CrossRef]
5. Asif, M. Health effects of omega-3,6,9 fatty acids: *Perilla frutescens* is a good example of plant oils. *Orient. Pharm. Exp. Med.* **2011**, *11*, 51–59. [CrossRef]
6. Arshad, M.S.; Batool, S.A. Natural Antimicrobials, their Sources and Food Safety. In *Food Additives*; Karunaratne, D.N., Pamunuwa, G., Eds.; IntechOpen: London, UK, 2017; pp. 87–102, ISBN 9781626239777.
7. Corrêa, R.C.G.; Di Gioia, F.; Ferreira, I.C.F.R.; Petropoulos, S.A. Wild greens used in the Mediterranean diet. In *The Mediterranean Diet: An Evidence-based Approach*; Preedy, V., Watson, R., Eds.; Academic Press: London, UK, 2020; pp. 209–228, ISBN 9788578110796.
8. Sodeifian, G.; Ardestani, N.S.; Sajadian, S.A.; Moghadamian, K. Properties of *Portulaca oleracea* seed oil via supercritical fluid extraction: Experimental and optimization. *J. Supercrit. Fluids* **2018**, *135*, 34–44. [CrossRef]
9. OECD/FAO. Oilseeds and oilseed products. In *OECD-FAO Agricultural Outlook 2016–2025*; OECD Publishing: Paris, France, 2016; pp. 127–138.
10. Petropoulos, S.A.; Fernandes, Â.; Calhelha, R.C.; Danalatos, N.; Barros, L.; Ferreira, I.C.F.R. How extraction method affects yield, fatty acids composition and bioactive properties of cardoon seed oil? *Ind. Crops Prod.* **2018**, *124*, 459–465. [CrossRef]
11. Petropoulos, S.A.; Fernandes, Â.; Arampatzis, D.A.; Tsiropoulos, N.G.; Petrović, J.; Soković, M.; Barros, L.; Ferreira, I.C.F.R. Seed oil and seed oil byproducts of common purslane (*Portulaca oleracea* L.): A new insight to plant-based sources rich in omega-3 fatty acids. *LWT Food Sci. Technol.* **2020**, *123*, 109099. [CrossRef]
12. Adarme-Vega, T.C.; Thomas-Hall, S.R.; Schenk, P.M. Towards sustainable sources for omega-3 fatty acids production. *Curr. Opin. Biotechnol.* **2014**, *26*, 14–18. [CrossRef]

13. Guil-Guerrero, J.L.; Gómez-Mercado, F.; Ramos-Bueno, R.P.; González-Fernández, M.J.; Urrestarazu, M.; Jiménez-Becker, S.; de Bélair, G. Fatty acid profiles and sn-2 fatty acid distribution of γ-linolenic acid-rich *Borago* species. *J. Food Compos. Anal.* **2018**, *66*, 74–80. [CrossRef]
14. Meddeb, W.; Rezig, L.; Zarrouk, A.; Nury, T.; Vejux, A.; Prost, M.; Bretillon, L.; Mejri, M.; Lizard, G. Cytoprotective activities of milk thistle seed oil used in traditional tunisian medicine on 7-ketocholesterol and 24S-hydroxycholesterol-induced toxicity on 158N murine oligodendrocytes. *Antioxidants* **2018**, *7*, 95. [CrossRef]
15. Siger, A.; Dwiecki, K.; Borzyszkowski, W.; Turski, M.; Rudzińska, M.; Nogala-Kałucka, M. Physicochemical characteristics of the cold-pressed oil obtained from seeds of *Fagus sylvatica* L. *Food Chem.* **2017**, *225*, 239–245. [CrossRef] [PubMed]
16. Kochhar, S.P. Sesame, rice-bran and flaxseed oils. In *Vegetable Oils in Food Technology: Composition, Properties and Uses*; Gunstone, F.D., Ed.; Blackwell Publishing Inc.: Boca Raton, FL, USA, 2002; pp. 297–326, ISBN 9780123849533.
17. Dubois, V.; Breton, S.; Linder, M.; Fanni, J.; Parmentier, M. Fatty acid profiles of 80 vegetable oils with regard to their nutritional potential. *Eur. J. Lipid Sci. Technol.* **2007**, *109*, 710–732. [CrossRef]
18. Wang, S.; Zhu, F.; Kakuda, Y. Sacha inchi (*Plukenetia volubilis* L.): Nutritional composition, biological activity, and uses. *Food Chem.* **2018**, *265*, 316–328. [CrossRef]
19. Knez Hrnčič, M.K.; Ivanovski, M.; Cör, D.; Knez, Ž. Chia Seeds (*Salvia hispanica* L.): An overview-phytochemical profile, isolation methods, and application. *Molecules* **2020**, *25*, 11. [CrossRef]
20. Tavarini, S.; Castagna, A.; Conte, G.; Foschi, L.; Sanmartin, C.; Incrocci, L.; Ranieri, A.; Serra, A.; Angelini, L.G. Evaluation of Chemical Composition of Two Linseed Varieties as Sources of Health-Beneficial Substances. *Molecules* **2019**, *24*, 3729. [CrossRef] [PubMed]
21. Dias, M.I.; Ferreira, I.C.F.R.; Barreiro, M.F. Microencapsulation of bioactives for food applications. *Food Funct.* **2015**, *6*, 1035–1052. [CrossRef]
22. Le Priol, L.; Gmur, J.; Dagmey, A.; Morandat, S.; El Kirat, K.; Saleh, K.; Nesterenko, A. Co-encapsulation of vegetable oils with phenolic antioxidants and evaluation of their oxidative stability under long-term storage conditions. *LWT Food Sci. Technol.* **2021**, *142*, 111033. [CrossRef]
23. Wrona, M.; Silva, F.; Salafranca, J.; Nerín, C.; Alfonso, M.J.; Caballero, M.Á. Design of new natural antioxidant active packaging: Screening flowsheet from pure essential oils and vegetable oils to ex vivo testing in meat samples. *Food Control* **2021**, *120*. [CrossRef]
24. Rabrenović, B.B.; Dimić, E.B.; Novaković, M.M.; Tešević, V.V.; Basić, Z.N. The most important bioactive components of cold pressed oil from different pumpkin (*Cucurbita pepo* L.) seeds. *LWT Food Sci. Technol.* **2014**, *55*, 521–527. [CrossRef]
25. Seymen, M.; Uslu, N.; Türkmen, Ö.; Al Juhaimi, F.; Özcan, M.M. Chemical Compositions and Mineral Contents of Some Hull-Less Pumpkin Seed and Oils. *JAOCS J. Am. Oil Chem. Soc.* **2016**, *93*, 1095–1099. [CrossRef]
26. Stevenson, D.G.; Eller, F.J.; Wang, L.; Jane, J.L.; Wang, T.; Inglett, G.E. Oil and tocopherol content and composition of pumpkin seed oil in 12 cultivars. *J. Agric. Food Chem.* **2007**, *55*, 4005–4013. [CrossRef]
27. Vujasinovic, V.; Djilas, S.; Dimic, E.; Romanic, R.; Takaci, A. Shelf life of cold-pressed pumpkin (*Cucurbita pepo* L.) seed oil obtained with a screw press. *JAOCS J. Am. Oil Chem. Soc.* **2010**, *87*, 1497–1505. [CrossRef]
28. Can-Cauich, C.A.; Sauri-Duch, E.; Moo-Huchin, V.M.; Betancur-Ancona, D.; Cuevas-Glory, L.F. Effect of extraction method and specie on the content of bioactive compounds and antioxidant activity of pumpkin oil from Yucatan, Mexico. *Food Chem.* **2019**, *285*, 186–193. [CrossRef]
29. Nederal, S.; Škevin, D.; Kraljić, K.; Obranović, M.; Papeša, S.; Bataljaku, A. Chemical composition and oxidative stability of roasted and cold pressed pumpkin seed oils. *JAOCS J. Am. Oil Chem. Soc.* **2012**, *89*, 1763–1770. [CrossRef]
30. Rezig, L.; Chouaibi, M.; Ojeda-Amador, R.M.; Gomez-Alonso, S.; Salvador, M.D.; Fregapane, G.; Hamdi, S. *Cucurbita maxima* pumpkin seed oil: From the chemical properties to the different extracting techniques. *Not. Bot. Horti Agrobot. Cluj-Napoca* **2018**, *46*, 663–669. [CrossRef]
31. Amin, M.Z.; Rity, T.I.; Uddin, M.R.; Rahman, M.M.; Uddin, M.J. A comparative assessment of anti-inflammatory, anti-oxidant and anti-bacterial activities of hybrid and indigenous varieties of pumpkin (*Cucurbita maxima* Linn.) seed oil. *Biocatal. Agric. Biotechnol.* **2020**, *28*, 101767. [CrossRef]
32. Obi, R.K.; Nwanebu, F.C.; Ndubuisi, U.U.; Orji, N.M. Antibacterial qualities and phytochemical screening of the oils of *Curcubita pepo* and *Brassica nigra*. *J. Med. Plants Res.* **2009**, *3*, 429–432.
33. Petropoulos, S.; Karkanis, A.; Fernandes, Â.; Barros, L.; Ferreira, I.C.F.R.; Ntatsi, G.; Petrotos, K.; Lykas, C.; Khah, E. Chemical composition and yield of six genotypes of common purslane (*Portulaca oleracea* L.): An alternative source of omega-3 fatty acids. *Plant Foods Hum. Nutr.* **2015**, *70*, 420–426. [CrossRef]
34. Guoyin, Z.; Hao, P.; Min, L.; Wei, G.; Zhe, C.; Changquan, L. Antihepatocarcinoma Effect of *Portulaca oleracea* L. in Mice by PI3K/Akt/mTOR and Nrf2/HO-1/NF- κ B Pathway. *Evid. Based Complement. Altern. Med.* **2017**, *2017*, 1–11. [CrossRef]
35. Eidi, A.; Mortazavi, P.; Moghadam, J.Z.; Mardani, P.M. Hepatoprotective effects of *Portulaca oleracea* extract against CCl 4 -induced damage in rats. *Pharm. Biol.* **2015**, *53*, 1042–1051. [CrossRef] [PubMed]
36. Egea-Gilabert, C.; Ruiz-Hernández, M.V.; Parra, M.Á.; Fernández, J.A. Characterization of purslane (*Portulaca oleracea* L.) accessions: Suitability as ready-to-eat product. *Sci. Hortic.* **2014**, *172*, 73–81. [CrossRef]
37. Gonnella, M.; National, I.; Charfeddine, M.; Agricultural, C.R.A.; Universit, G.C.; Universit, P.S.; Moro, B.A. Purslane: A review of its potential for health and agricultural aspects. *Eur. J. Plant Sci. Biotechnol.* **2010**, *4*, 131–136.

38. Alvarado-Gómez, E.; Tapia, J.I.; Encinas, A. A sustainable hydrophobic luffa sponge for efficient removal of oils from water. *Sustain. Mater. Technol.* **2021**, *28*, e00273. [CrossRef]
39. Adewuyi, A.; Oderinde, R.A.; Rao, B.V.S.K.; Prasad, R.B.N.; Anjaneyulu, B. *Blighia unijugata* and *Luffa cylindrica* Seed Oils: Renewable Sources of Energy for Sustainable Development in Rural Africa. *Bioenergy Res.* **2012**, *5*, 713–718. [CrossRef]
40. Ali, M.A.; Azad, M.A.K.; Yeasmin, M.S.; Khan, A.M.; Sayeed, M.A. Oil characteristics and nutritional composition of the ridge gourd (*Luffa acutangula* Roxb.) seeds grown in Bangladesh. *Food Sci. Technol. Int.* **2009**, *15*, 243–250. [CrossRef]
41. Kamel, B.S.; Blackman, B. Nutritional and oil characteristics of the seeds of angled Luffa *Luffa acutangula*. *Food Chem.* **1982**, *9*, 277–282. [CrossRef]
42. Muthumani, P.; Meera, R.; Mary, S.; Jeenamathew; Devi, P.; Kameswari, B.; Eswara Priya, B. Phytochemical screening and anti inflammatory, bronchodilator and antimicrobial activities of the seeds of *Luffa cylindrica*. *Res. J. Pharm. Biol. Chem. Sci.* **2010**, *1*, 11–22.
43. Spréa, R.M.; Fernandes, Â.; Calhelha, R.C.; Pereira, C.; Pires, T.C.S.P.; Alves, M.J.; Canan, C.; Barros, L.; Amaral, J.S.; Ferreira, I.C.F.R. Chemical and bioactive characterization of the aromatic plant *Levisticum officinale* W.D.J. Koch: A comprehensive study. *Food Funct.* **2020**, *11*, 1292–1303. [CrossRef] [PubMed]
44. Abreu, R.M.V.; Ferreira, I.C.F.R.; Calhelha, R.C.; Lima, R.T.; Vasconcelos, M.H.; Adega, F.; Chaves, R.; Queiroz, M.J.R.P. Anti-hepatocellular carcinoma activity using human HepG2 cells and hepatotoxicity of 6-substituted methyl 3-aminothieno[3,2-b]pyridine-2- carboxylate derivatives: In vitro evaluation, cell cycle analysis and QSAR studies. *Eur. J. Med. Chem.* **2011**, *46*, 5800–5806. [CrossRef] [PubMed]
45. Finimundy, T.C.; Karkanis, A.; Fernandes, Â.; Petropoulos, S.A.; Calhelha, R.; Petrović, J.; Soković, M.; Rosa, E.; Barros, L.; Ferreira, I.C.F.R. Bioactive properties of *Sanguisorba minor* L. cultivated in central Greece under different fertilization regimes. *Food Chem.* **2020**, *327*, 127043. [CrossRef]
46. Zamani Ghaleshahi, A.; Ezzatpanah, H.; Rajabzadeh, G.; Ghavami, M. Comparison and analysis characteristics of flax, perilla and basil seed oils cultivated in Iran. *J. Food Sci. Technol.* **2020**, *57*, 1258–1268. [CrossRef]
47. Ren, G.; Zhang, W.; Sun, S.; Duan, X.; Zhang, Z. Enhanced extraction of oil from flaxseed (*Linum usitatissimum* L.) using microwave pre-treatment. *J. Oleo Sci.* **2015**, *64*, 1043–1047. [CrossRef]
48. Gutte, K.B.; Sahoo, A.K.; Ranveer, R.C. Effect of ultrasonic treatment on extraction and fatty acid profile of flaxseed oil. *OCL Oilseeds Fats* **2015**, *22*. [CrossRef]
49. Liu, L.; Howe, P.; Zhou, Y.F.; Xu, Z.Q.; Hocart, C.; Zhang, R. Fatty acids and β-carotene in Australian purslane (*Portulaca oleracea*) varieties. *J. Chromatogr. A* **2000**, *893*, 207–213. [CrossRef]
50. Delfan-Hosseini, S.; Nayebzadeh, K.; Mirmoghtadaie, L.; Kavosi, M.; Hosseini, S.M. Effect of extraction process on composition, oxidative stability and rheological properties of purslane seed oil. *Food Chem.* **2017**, *222*, 61–66. [CrossRef] [PubMed]
51. Ionescu, M.; Voicu, G.; Sorin-Stefan, B.; Covaliu, C.; Dincă, M.; Ungureanu, N. Parameters influencing the screw pressing process of oilseed materials. In Proceedings of the 2nd International Conference on Thermal Equipment, Renewable Energy and Rural Development, Băile Olăneşti, Romania, 20–22 July 2013; pp. 243–248.
52. Nyam, K.L.; Tan, C.P.; Lai, O.M.; Long, K.; Che Man, Y.B. Physicochemical properties and bioactive compounds of selected seed oils. *LWT Food Sci. Technol.* **2009**, *42*, 1396–1403. [CrossRef]
53. Murkovic, M.; Hillebrand, A.; Winkler, J.; Leitner, E.; Pfannhauser, W. Variability of fatty acid content in pumpkin seeds (*Cucurbita pepo* L.). *Eur. Food Res. Technol.* **1996**, *203*, 216–219. [CrossRef] [PubMed]
54. García-Díaz, M.T.; Martínez-Rivas, J.M.; Mancha, M. Temperature and oxygen regulation of oleate desaturation in developing sunflower (*Helianthus annuus*) seeds. *Physiol. Plant.* **2002**, *114*, 13–20. [CrossRef]
55. Guil, J.L.; Torija, M.E.; Giménez, J.J.; Rodriguez, I. Identification of fatty acids in edible wild plants by gas chromatography. *J. Chromatogr. A* **1996**, *719*, 229–235. [CrossRef]
56. Petropoulos, S.A.; Fernandes, Â.; Dias, M.I.; Pereira, C.; Calhelha, R.C.; Chrysargyris, A.; Tzortzakis, N.; Ivanov, M.; Sokovic, M.D.; Barros, L.; et al. Chemical composition and plant growth of *Centaurea raphanina* subsp. *mixta* plants cultivated under saline conditions. *Molecules* **2020**, *25*, 2204. [CrossRef]
57. Petropoulos, S.; Fernandes, Â.; Karkanis, A.; Ntatsi, G.; Barros, L.; Ferreira, I. Successive harvesting affects yield, chemical composition and antioxidant activity of *Cichorium spinosum* L. *Food Chem.* **2017**, *237*, 83–90. [CrossRef]
58. Yoon, B.K.; Jackman, J.A.; Valle-González, E.R.; Cho, N.J. Antibacterial Free Fatty Acids and Monoglycerides: Biological Activities, Experimental Testing, and Therapeutic Applications. *Int. J. Mol. Sci.* **2018**, *19*, 1114. [CrossRef]
59. Dashora, N.; Chauhan, L.S.; Kumar, N. In vitro Cytotoxic Activity of *Luffa acutangula* on Human Neuronal Glioblastoma and Human Lung Adenocarcinoma Cell Lines. *Sch. Acad. J. Pharm.* **2014**, *3*, 401–405.
60. Vanajothi, R.; Sudha, A.; Manikandan, R.; Rameshthangam, P.; Srinivasan, P. *Luffa acutangula* and *Lippia nodiflora* leaf extract induces growth inhibitory effect through induction of apoptosis on human lung cancer cell line. *Biomed. Prev. Nutr.* **2012**, *2*, 287–293. [CrossRef]
61. Dashora, N.; Chauhan, L.S. Evaluation of antitumor potential of *Luffa acutangula* on Ehrlich's ascites carcinoma treated mice. *Int. J. Pharm. Clin. Res.* **2015**, *7*, 296–299.
62. Shendge, P.N.; Belemkar, S. Therapeutic potential of *Luffa acutangula*: A review on its traditional uses, phytochemistry, pharmacology and toxicological aspects. *Front. Pharmacol.* **2018**, *9*. [CrossRef] [PubMed]

63. Mason, J.K.; Chen, J.; Thompson, L.U. Flaxseed oil-trastuzumab interaction in breast cancer. *Food Chem. Toxicol.* **2010**, *48*, 2223–2226. [CrossRef]
64. Sorice, A.; Guerriero, E.; Volpe, M.G.; Capone, F.; La Cara, F.; Ciliberto, G.; Colonna, G.; Costantini, S.; McPhee, D.J. Differential response of two human breast cancer cell lines to the phenolic extract from flaxseed oil. *Molecules* **2016**, *21*, 319. [CrossRef]
65. Truan, J.S.; Chen, J.M.; Thompson, L.U. Flaxseed oil reduces the growth of human breast tumors (MCF-7) at high levels of circulating estrogen. *Mol. Nutr. Food Res.* **2010**, *54*, 1414–1421. [CrossRef] [PubMed]
66. Medjakovic, S.; Hobiger, S.; Ardjomand-woelkart, K.; Bucar, F.; Jungbauer, A. Pumpkin seed extract: Cell growth inhibition of hyperplastic and cancer cells, independent of steroid hormone receptors. *Fitoterapia* **2016**, *110*, 150–156. [CrossRef]
67. Chari, K.Y.; Polu, P.R.; Shenoy, R.R. An Appraisal of Pumpkin Seed Extract in 1, 2-Dimethylhydrazine Induced Colon Cancer in Wistar Rats. *J. Toxicol.* **2018**, *2018*. [CrossRef]
68. Abou-Elella, F.; Mourad, R. Anticancer and anti-oxidant potentials of ethanolic extracts of *Phoenix dactylifera*, *Musa acuminata* and *Cucurbita maxima*. *Res. J. Pharm. Biol. Chem. Sci.* **2015**, *6*, 710–720.
69. Colagar, A.H.; Souraki, O.A. Review of Pumpkin Anticancer Effects. *Quran Med.* **2011**, *1*, 77–88. [CrossRef]
70. Heng, C.X.; Li, F.; Li, Z.; Zhang, Z.C. Purification and characterization of Moschatin, a novel type I ribosome-inactivating protein from the mature seeds of pumpkin (*Cucurbita moschata*), and preparation of its immunotoxin against human melanoma cells. *Cell Res.* **2003**, *13*, 369–374. [CrossRef]
71. Khan, N.; Jajeh, F.; Khan, M.I.; Mukhtar, E.; Shabana, S.M.; Mukhtar, H. Sestrin-3 modulation is essential for therapeutic efficacy of cucurbitacin B in lung cancer cells. *Carcinogenesis* **2017**, *38*, 184–195. [CrossRef]
72. Bardaa, S.; Halima, N.B.; Aloui, F.; Mansour, R.B.; Jabeur, H.; Bouaziz, M. Oil from pumpkin (*Cucurbita pepo* L.) seeds: Evaluation of its functional properties on wound healing in rats. *Lipids Health Dis.* **2016**, *15*, 1–12. [CrossRef]
73. Al-Okbi, S.Y.; Mohamed, D.A.; Kandil, E.; Ahmed, E.K.; Mohammed, S.E. Antioxidant and anti-cancer effect of Egyptian and European pumpkin seed oil. *Res. J. Pharm. Biol. Chem. Sci.* **2016**, *7*, 574–580.
74. Silva, R.; Carvalho, I.S. In vitro antioxidant activity, phenolic compounds and protective effect against DNA damage provided by leaves, stems and flowers of *Portulaca oleracea* (Purslane). *Nat. Prod. Commun.* **2014**, *9*, 45–50. [CrossRef] [PubMed]
75. Tian, J.L.; Liang, X.; Gao, P.Y.; Li, D.Q.; Sun, Q.; Li, L.Z.; Song, S.J. Two new alkaloids from *Portulaca oleracea* and their cytotoxic activities. *J. Asian Nat. Prod. Res.* **2014**, *16*, 259–264. [CrossRef]
76. Al-Sheddi, E.S.; Farshori, N.N.; Al-Oqail, M.M.; Musarrat, J.; Al-Khedhairy, A.A.; Siddiqui, M.A. *Portulaca oleracea* seed oil exerts cytotoxic effects on human liver cancer (HepG2) and human lung cancer (A-549) cell lines. *Asian Pac. J. Cancer Prev.* **2015**, *16*, 3383–3387. [CrossRef]
77. Al-Mathkhury, H.J.F.; Al-Dhamin, A.S.; Al-Taie, K.L. Antibacterial and Antibiofilm Activity of Flaxseed Oil. *Iraqi J. Sci.* **2016**, *57*, 1086–1095.
78. Kaithwas, G.; Mukerjee, A.; Kumar, P.; Majumdar, D.K. *Linum usitatissimum* (linseed/flaxseed) fixed oil: Antimicrobial activity and efficacy in bovine mastitis. *Inflammopharmacology* **2011**, *19*, 45–52. [CrossRef] [PubMed]
79. Santos, J.S.; Escher, G.B.; da Silva Pereira, J.M.; Marinho, M.T.; Prado-Silva, L.d.; Sant'Ana, A.S.; Dutra, L.M.; Barison, A.; Granato, D. 1H NMR combined with chemometrics tools for rapid characterization of edible oils and their biological properties. *Ind. Crops Prod.* **2018**, *116*, 191–200. [CrossRef]
80. Joshi, Y.; Garg, R.; Juyal, D. Evaluation of synergistic antimicrobial activity of Gemifloxacin with *Linum usitatissimum* seed oil. *J. Phytopharm.* **2014**, *3*, 384–388.
81. Swain, T.; Sahoo, R.K.; Kar, D.M.; Subudhi, E. Phytomedicinal potential of *Luffa cylindrica* (L.) Reom extracts. *J. Pure Appl. Microbiol.* **2013**, *7*, 697–703.
82. Ercisli, S.; Coruh, I.; Gormez, A.; Sengul, M. Antioxidant and antibacterial activities of *Portulaca oleracea* L. grown wild in Turkey. *Ital. J. Food Sci.* **2008**, *20*, 533–542.
83. Oraibi, A.; AlShammari, A.; Mohsien, R.; Obaid, W. Investigation the Antibacterial Activity of *Portulaca oleracea* L. Tissue Cultures in vitro. *J. Pharm. Res. Int.* **2017**, *18*, 1–7. [CrossRef]
84. Rahimi, V.B.; Ajam, F.; Rakhshandeh, H.; Askari, V.R. A pharmacological review on *Portulaca oleracea* L.: Focusing on anti-inflammatory, anti- oxidant, immuno-modulatory and antitumor activities. *J. Pharmacopunct.* **2019**, *22*, 7–15. [CrossRef]
85. Du, Y.K.; Liu, J.; Li, X.M.; Pan, F.F.; Wen, Z.G.; Zhang, T.C.; Yang, P.L. Flavonoids extract from *Portulaca oleracea* L. induce *Staphylococcus aureus* death by apoptosis-like pathway. *Int. J. Food Prop.* **2017**, *20*, S534–S542. [CrossRef]
86. Tayel, A.A.; Shaban, S.M.; Moussa, S.H.; Elguindy, N.M.; Diab, A.M.; Mazrou, K.E.; Ghanem, R.A.; El-Sabbagh, S.M. Bioactivity and application of plant seeds' extracts to fight resistant strains of Staphylococcus aureus. *Ann. Agric. Sci.* **2018**, *63*, 47–53. [CrossRef]
87. Bakkiyaraj, S.; Pandiyaraj, S. Evaluation of potential antimicrobial activity of some medicinal plants against common food-borne pathogenic microorganism. *Int. J. Pharma Bio Sci.* **2011**, *2*, 484–491.
88. Othman, A.S. Bactericidal efficacy of omega-3 fatty acids and esters present in moringa oleifera and Portulaca oleracea fixed oils against oral and gastro enteric bacteria. *Int. J. Pharmacol.* **2017**, *13*, 44–53. [CrossRef]
89. Hammer, K.A.; Carson, C.F.; Riley, T. V Antimicrobial activity of essential oils and other plant extracts. *J. Appl. Microbiol.* **1999**, *86*, 985–990. [CrossRef] [PubMed]
90. Xuan, T.D.; Gangqiang, G.; Minh, T.N.; Quy, T.N.; Khanh, T.D. An overview of chemical profiles, antioxidant and antimicrobial activities of commercial vegetable edible oils marketed in Japan. *Foods* **2018**, *7*, 21. [CrossRef]

91. Yang, R.; Wang, S.; Zhang, L.; Wang, X.; Ma, F.; Mao, J.; Zhang, Q. Evaluation and comparison of antibacterial activities of edible vegetable oils in China. *Oil Crop Sci.* **2018**, *3*, 57–62. [CrossRef]
92. Abdelillah, A.; Houcine, B.; Halima, D.; Meriel, C.S.; Imane, Z.; Eddine, S.D.; Abdallah, M.; Daoudi, C.S. Evaluation of antifungal activity of free fatty acids methyl esters fraction isolated from Algerian *Linum usitatissimum* L. seeds against toxigenic Aspergillus. *Asian Pac. J. Trop. Biomed.* **2013**, *3*, 443–448. [CrossRef]
93. Amrithaa, B.; Geetha, R.V. Comparative evaluation of antimycotic activity of *Cucurbita maxima* seed oil and orange peel oil in reducing *Candida albicans* count. *Drug Invent. Today* **2020**, *14*, 1210–1213.
94. Abd El-Aziz, A.B.; Abd El-Kalek, H.H. Antimicrobial proteins and oil seeds from pumpkin (*Cucurbita moschata*). *Nat. Sci.* **2011**, *9*, 105–119.
95. Sener, B.; Orhan, I.; Ozcelik, B.; Kartal, M.; Aslan, S.; Ozbilen, G. Antimicrobial and Antiviral Activities of Two Seed Oil. *Nat. Prod. Commun.* **2007**, *2*, 395–398.

Article

Effect of Photosensitization Mediated by Curcumin on Carotenoid and Aflatoxin Content in Different Maize Varieties

Rafael Nguenha [1,2], Maral Seidi Damyeh [3,4], Anh D. T. Phan [3,4], Hung T. Hong [4], Mridusmita Chaliha [3,4], Tim J. O'Hare [4], Michael E. Netzel [3,4] and Yasmina Sultanbawa [3,4,*]

1 School of Agriculture and Food Sciences, The University of Queensland, St. Lucia, QLD 4072, Australia; r.nguenha@uq.edu.au
2 Faculdade de Agronomia e Engenharia Florestal, Universidade Eduardo Mondlane, Av. Julius Nyerere 3453, Maputo 1102, Mozambique
3 ARC Industrial Transformation Training Centre for Uniquely Australian Foods, Queensland Alliance for Agriculture and Food Innovation, The University of Queensland, Coopers Plains, QLD 4108, Australia; s.maral@uq.edu.au (M.S.D.); anh.phan1@uq.net.au (A.D.T.P.); m.chaliha@uq.edu.au (M.C.); m.netzel@uq.edu.au (M.E.N.)
4 Centre for Nutrition and Food Sciences, Queensland Alliance for Agriculture and Food Innovation, The University of Queensland, St. Lucia, QLD 4072, Australia; h.trieu@uq.edu.au (H.T.H.); t.ohare@uq.edu.au (T.J.O.)
* Correspondence: y.sultanbawa@uq.edu.au

Citation: Nguenha, R.; Damyeh, M.S.; Phan, A.D.T.; Hong, H.T.; Chaliha, M.; O'Hare, T.J.; Netzel, M.E.; Sultanbawa, Y. Effect of Photosensitization Mediated by Curcumin on Carotenoid and Aflatoxin Content in Different Maize Varieties. *Appl. Sci.* **2021**, *11*, 5902. https://doi.org/10.3390/app11135902

Academic Editors: Gwiazdowska Daniela, Krzysztof Juś and Katarzyna Marchwińska

Received: 13 May 2021
Accepted: 23 June 2021
Published: 25 June 2021

Abstract: Mycotoxins are naturally occurring toxins produced by certain types of fungi that contaminate food and feed, posing serious health risks to human and livestock. This study evaluated the combination of blue light with curcumin to inactivate *Aspergillus flavus* spores, its effect on aflatoxin B1 (AFB1) production and maintaining carotenoid content in three maize varieties. The study was first conducted in vitro, and the spore suspensions (10^4 CFU·mL^{-1}) were treated with four curcumin concentrations (25 and 50 μM in ethanol, 1000 and 1250 μM in propylene glycol) and illuminated at different light doses from 0 to 130.3 J·cm^{-2}. The photoinactivation efficiency was light-dose dependent with the highest photoinactivation of 2.3 log CFU·mL^{-1} achieved using 1000 μM curcumin at 104.2 J·cm^{-2}. Scanning electron microscopy revealed cell wall deformations as well as less density in photosensitized cells. Photosensitization of maize kernels gave rise to a complete reduction in the viability of *A. flavus* and therefore inhibition of AFB1 production, while no significant ($p > 0.05$) effect was observed using either light or curcumin. Moreover, photosensitization did not affect the carotenoids in all the studied maize varieties. The results suggest that photosensitization is a green alternative preservation technique to decontaminate maize kernels and reduce consumer exposure to AFB1 without any effect on carotenoid content.

Keywords: aflatoxin B1; *Aspergillus flavus*; carotenoids; curcumin; maize varieties; photosensitization; preservation technique

1. Introduction

Maize (*Zea mays* L.) is one of the most important cereal crops in the world, coming third after wheat and rice. This crop can be used for several purposes such as animal feed, raw material in the industry and human consumption. Nevertheless, maize is highly susceptible to contamination by fungi. Depending on the environmental and storage conditions, fungi may produce mycotoxins, which are toxic secondary metabolites contaminating foods throughout the value chain [1]. Contamination of maize by mycotoxin is of serious concern, especially in developing countries where maize is a staple food. Owing to their genotoxic and carcinogenic effect, aflatoxins produced by *Aspergillus flavus* Link and *Aspergillus parasiticus* Speare are classified as the most poisonous mycotoxins [2]. Normal food processing techniques such as heat treatment and freezing are not sufficient to destroy mycotoxins [3]. Thus, there is a need for research in novel preservation techniques to control fungal growth in food and feed and prevent mycotoxin accumulation, particularly green alternatives to the current conventional preservation techniques.

Photosensitization, a technique that uses a photoactive compound (photosensitizer) to eradicate microorganisms by the generation of reactive oxygen species (ROS), was recently introduced as an alternative approach to decontaminate food products and food contact surfaces [4]. This technique has shown promising inhibitory effects against yeasts, mould, viruses, parasites, Gram-negative and Gram-positive bacteria both in vitro and in vivo [5], mainly depositing on the surface of fruits and fresh-cut vegetables [6].

Curcumin is a predominant polyphenolic compound in turmeric (*Curcuma longa* L.) rhizomes that has been extensively used for several years as a natural food colorant and dietary supplement [7]. Recent studies have demonstrated the phototoxic activity of curcumin when combined with blue light in some food materials such as maize, fresh date fruit and oysters [8,9]. Curcumin presents a maxima absorption intensity in the region between 414 and 434 nm, depending on the solvent [10]. In the food matrix, curcumin absorption range can overlap with naturally occurring substances, such as yellow carotenoids which are known to absorb light in a broad range from 400 to 500 nm, with three maxima and shoulders [11,12].

Carotenoids are naturally occurring pigments that are abundant in plants. Consumption of carotenoid-rich food is associated with a low incidence of several diseases such as cardiovascular diseases, cancers, age-related macular degeneration, and cataracts [13]. Additionally, there is substantial evidence suggesting a positive effect of high carotenoid concentration on inhibiting aflatoxin biosynthesis [14–16]. Nevertheless, carotenoids are light-sensitive and can degrade into other compounds or isomerise into different conformation via photo-isomerisation [17], which could lead to the loss of their biological function. Moreover, carotenoids are singlet oxygen quencher [18,19], and previous studies have shown that β-carotene could prevent photosensitized oxidation [20,21]. Therefore, high concentration of carotenoids in food could potentially affect the efficiency of singlet oxygen generation during photosensitization, resulting in lower microbial photoinactivation.

Information regarding the effect of curcumin-mediated photosensitization on the nutritional profile and sensory properties of food is limited, especially for food grains including maize. Temba et al. [22] evaluated the effect of curcumin-mediated photoinactivation on *A. flavus* spores in maize kernels and flour, achieving 2 log CFU·g^{-1} reduction. In another study, curcumin-mediated photosensitization could reduce 73% AFB1 concentration in maize kernels after 10 days of storage at 26 °C [23]. Glueck et al. [4] reported up to 4 log CFU·g^{-1} reduction of *Escherichia coli* on mung beans and fenugreek seeds, using 435-nm LED array in combination with SACUR-3, a curcumin derivative. Nevertheless, none of these studies evaluated the effect of photosensitization on the nutritional profile and sensory properties of the studied food material. Therefore, this study aimed to evaluate the effect of photosensitization using LEDs and curcumin on inactivation of *A. flavus* growth and aflatoxin B1 accumulation on three maize varieties, as well as its influence on maize kernel colour, individual carotenoid compounds and total carotenoid content.

2. Materials and Methods

2.1. Inoculum Preparation

An aflatoxin producing reference strain of *A. flavus* ATCC 26944 (American Type Culture Collection, In Vitro Technologies Pty. Ltd., Noble Park, VIC, Australia) was grown on malt extract agar (MEA) (Thermo Fisher Scientific Pty. Ltd., Scoresby, VIC, Australia) for seven days at 25 °C. Spore harvest was done by flooding the culture surface with 10 mL of phosphate buffer saline (PBS) solution containing 0.1% Tween 80 (polyoxyethylene (20) sorbitan monooleate), followed by gentle swirling. The suspension was then collected and vortexed for one minute, followed by centrifuging (Eppendorf Centrifuge 5415 D, Hamburg-Eppendorf, Germany) at 10,000 rpm for another minute to sediment the spores. The supernatant was removed, the pellet was resuspended in sterile PBS containing 0.1%

Tween 80. Then, 15% (v/v) of glycerol was added to the suspension and the spore solution was stored at $-20\ ^{\circ}C$ until further experiments.

2.2. Photosensitizer Solutions

Curcumin (Sigma Aldrich, St. Louis, MO, USA) stock solution (2000 µM) was prepared using absolute ethanol (≥99.8%; Sigma Aldrich) or 50% (v/v) aqueous propylene glycol (≥99.5%; Sigma Aldrich). The working solutions were freshly prepared prior to the photosensitization by diluting the curcumin stock solutions with distilled water to different concentrations at 25 and 50 µM for ethanol, and 1000 and 1250 µM for propylene glycol. Curcumin concentrations were defined based on previous experiments using a xenon arc lamp, which has shown that when dissolved in ethanol curcumin is effective at concentrations below 100 µM, while when dissolved in propylene glycol it is effective at concentrations above 750 µM [22,24].

2.3. Light Source

High-intensity 430 ± 3 nm LED was used for photosensitization treatment. Temperature and relative humidity inside the light-box were monitored using Inkbird Bluetooth mini smart sensor (IBS-TH1 mini, Inkbird, Regents Park, NSW, Australia) (±0.3 °C temp., ±3% RH). The LED was accommodated in a box impermeable to light, with built in cooling fans to prevent the heat accumulation during illumination. The photosynthetic photon flux density (PPFD) was monitored during illumination using a spectrometer (HR-450, Hipoint, Taichung City, Taiwan). The PPFD ($\mu mol \cdot m^{-2} \cdot s^{-1}$) was converted to irradiance ($W \cdot cm^{-2}$) to calculate the light dose delivered to the sample after a particular time of illumination using Equation (1) [25]:

$$E = Pt \quad (1)$$

where E is the light dose ($J \cdot cm^{-2}$), P is the irradiance ($W \cdot cm^{-2}$), and t is the illumination time (s).

2.4. Maize Kernel Samples

Three maize varieties including yellow (Pioneer® hybrid P1756), white (Pioneer® hybrid P1477W), and popcorn maize accession were used for in vivo trials. White and yellow maize kernels were bought from Pioneer® Seeds Australia (Brisbane, QLD, Australia), and the popcorn maize accession was obtained from the University of Queensland's breeding program. All samples were kept at room temperature and used within 3 months.

2.5. In Vitro Trials-Effect of Photosensitization on A. flavus Spores

2.5.1. Study Design

Four concentrations of curcumin solutions (stated in Section 2.2) were combined with eight different light doses, ranging from 0 to 130.3 J/cm^2. Four applied treatments included photosensitization (the combination of light and curcumin-P^+L^+), the negative control (no curcumin, no light-P^oL^o), light treatment (light, no curcumin-P^oL^+), and curcumin treatment (curcumin, no light-P^+L^o). Ethanol and propylene glycol at concentrations up to 2.5% (v/v) and 25% (v/v), respectively, were also included to test any effects from the solvents on *A. flavus* growth using an agar well diffusion method [26]. Three sets of the experiments were conducted, and all treatments were conducted in triplicate in each set.

2.5.2. Evaluation of Photosensitization Effect on A. flavus Spores

For photosensitization trials, 1 mL of *A. flavus* spore suspension (equivalent to 10^4 $CFU \cdot mL^{-1}$) was mixed with 1 mL of curcumin solution in a small Petri dish (35 mm × 10 mm), followed by incubation for 10 min at room temperature in dark before illumination. Peptone water (0.1%; w/v) was used instead of photosensitizer in the negative control (P^oL^o) and light treatment (P^oL^+). Then, the Petri dish (without lid) was placed under the light source with 10 cm distance and illuminated at 430 ± 3 nm. After treatment, 100 µL aliquots of the photosensitized mixtures were plated onto MEA and incubated at

25 °C for 72 h to determine the Colony-Forming Unit (CFU) using a colony counter (Stuart Scientific, Stone, UK). The fungal survival was calculated by log reduction (LR) of CFU following Equation (2):

$$LR = Log_{10}(A) - Log_{10}(B) \quad (2)$$

where: A: CFU·mL^{-1} of the negative control; B: CFU·mL^{-1} of photosensitized spores.

2.5.3. Determination of Reactive Oxygen Species (ROS) Formation

The generation of singlet oxygen was monitored using 1,3-diphenylisobenzofuran (DPBF; Arcos Organics, Thermo Fisher Scientific Pty. Ltd., Scoresby, VIC, Australia) as a specific reactant [27,28]. Briefly, equal aliquots (100 μL) of photosensitizer and DPBF solutions (50 μg/mL prepared in absolute ethanol) were mixed and illuminated at the investigated light doses. DPBF absorbance was monitored at 410 nm using a micro-plate reader (Infinite® M200, Tecan, Port Melbourne, VIC, Australia). The absorbance of DPBF was monitored every 10 min for a total duration of 30 min, to obtain an estimate about the time-dependent production of ROS after the photosensitizer was subjected to light illumination. The rate of ROS production is proportional to the rate of decrease of DPBF absorbance at 410 nm as a function of illumination time, represented by the slope of the equation.

2.5.4. Morphological Changes on *A. flavus* Spores

The scanning electron microscopy (SEM) was used to evaluate the morphology of *A. flavus* spores subjected to different treatment conditions. Sample preparation was carried out according to Smijs et al. [29] with some modifications. Inoculum of both photosensitized spores (~10^5 CFU·mL^{-1}) and control (untreated spores) were firstly centrifuged at 10,000 rpm for 15 min to remove the media, then washed three times with 0.1 M PBS, followed by centrifugation to remove the unbound photosensitizer. The pellet was mixed with 3% glutaraldehyde (prepared in 0.1 M sodium cacodylate buffer at pH 7.2) and incubated at room temperature for 1 h, then at 4 °C overnight. Next, the samples were dried overnight at room temperature before observation by SEM (Neoscope JCM-5000 SEM, JEOL Ltd., Tokyo, Japan).

2.6. In Vivo Trials-Effect of Photosensitization on Maize Kernels

2.6.1. Inoculation of Kernels

Dry maize kernels were sterilized by autoclaving (121 °C, 15 psi, 15 min). Three replicates of 1.5 g of sterile maize were immersed in 5 mL of sterile PBS, followed by mixing at 1400 rpm for 30 s using a vortex. From the resulting suspension, 100 μL aliquot was plated on MEA to determine CFU of negative control. For positive control, the remaining sterilized maize was inoculated by immersing in 500 mL spore's suspension (10^5 CFU·mL^{-1}) for 30 s, the liquid was decanted, and the maize kernels were dried at 25 °C overnight, followed by CFU determination as described for negative control.

2.6.2. Photosensitization Treatment

The inoculated maize kernels were immersed in 1000 μM curcumin (dissolved in propylene glycol) for 5 min under dim light/dark. The kernels were then removed from curcumin solution and illuminated at 430 nm using a light dose of 104.2 J/cm^2. In the control treatment, curcumin solution was replaced by PBS. After illumination, the kernels were dried at 25 °C overnight, and ca. 1.5 g of dried kernels were mixed with PBS (5 mL), vortexed at 1400 rpm for 30 s, and plated on MEA to determine the CFU.

2.7. Extraction and Quantification of Aflatoxin B1

After the photosensitization treatment, Petri dishes (3 cm diameter) containing maize samples were placed inside 6 cm-diameter Petri dishes filled with water to provide favourable growth conditions for the mould, and the maize samples were incubated for 14 days at 25 ± 0.5 °C. AFB1 was determined at 7-day intervals. AFB1 content was determined as previously reported by Suylok et al. [30], with some modifications. Briefly, extraction solvent (2 mL) of acetonitrile/Milli-Q water/formic acid at a ratio of 79/20/1 (*v*/*v*/*v*) was added to 0.5 g of maize kernels. The mixture was mechanically shaken for 90 min using an orbital shaker (Ratek OM1, Ratek Instruments Pty Ltd., Boronia, VIC, Australia), and subsequently centrifuged (Eppendorf Centrifuge 5810 R) at 3000 rpm for 2 min at room temperature. The supernatant (1 mL) was diluted with an equal volume of the extraction solvent and filtered through a 0.22 μm hydrophilic PTFE syringe filter into a HPLC vial for AFB1 analysis by a Shimadzu LC-ESI-MS/MS (Shimadzu, Kyoto, Japan).

2.8. Effect of Photosensitization on Maize Kernel Colour

Potential changes in the colour of photosensitized maize kernels were determined using a handheld colorimeter (Konica Minolta CR-400, Thermo Fisher Scientific Pty Ltd., Scoresby, VIC, Australia). The colour parameters L*, a*, and b* were measured in three replicates (four kernels per replication). The total colour change was calculated using Equation (3). Non-treated kernels were used as a reference.

$$\Delta E = (\Delta L^{*2} + \Delta a^{*2} + \Delta b^{*2})^{1/2} \quad (3)$$

Chroma and hue angle were calculated using Equations (4) and (5):

$$\text{Chroma} = [(a^*)^2 + (b^*)^2]^{1/2} \quad (4)$$

$$\text{Hue} (^\circ) = \tan^{-1} (b^*/a^*) \quad (5)$$

The presence of residual curcumin inside the maize kernel was evaluated using a spectrophotometer (Genesys^{TM} 20, Thermo Fisher Scientific Pty Ltd., Scoresby, VIC, Australia). Five grams of the kernels were milled into a fine powder using a coffee grinder and the powder samples were extracted with 4 mL of 50% propylene glycol (*v*/*v*), followed by centrifugation at 10,000 rpm for 5 min. Curcumin solution in propylene glycol (750 μM) was included as a reference, and maize kernels that were not soaked in curcumin were also used as a blank. The absorption intensity of the extract was monitored from 390 nm to 700 nm using a micro-plate reader.

2.9. Carotenoid Extraction and Quantification

The extraction and quantification of carotenoids were conducted following the procedures of Saini and Keum [31] with some modifications. Briefly, maize kernels were first ground using a coffee grinder and then milled with a ball mill (Retsch MM 301, Metrohm, Brendale, QLD, Australia) to obtain a fine powder. Powdered maize samples (0.5 g) were mixed with 6 mL of ethanol containing 0.1% butylated hydroxytoluene (BHT) (*w*/*v*) and vortexed for 30 s. Ten millilitres (mL) solvent mixture of DCM and hexane (30:70; *v*/*v*) containing 0.1% BHT was added to the mixture to extract carotenoid compounds into the upper layer. NaCl (3 mL, 5%) was added to the mixture to facilitate phase separation, followed by centrifugation (Eppendorf Centrifuge 5810 R) at 3900 rpm for 10 min. Then, the supernatant was collected, and the pellet was re-extracted twice with DCM/hexane mixture with sonication for 10 min at 25 ± 2 °C. All extraction steps were performed under dim light to prevent the isomerization of carotenoids. The supernatants were combined and evaporated until dryness using nitrogen gas. The dried extract was redissolved in methyl tert-butyl ether (MTBE) and methanol (1:1, *v*/*v*) solution containing 0.1% BHT and filtered through a 0.22 μm hydrophilic PTFE syringe filter before UPLC-PDA analysis.

2.10. Data Analysis

The results are reported as mean values with standard errors (SE). Data analysis was conducted using Microsoft Excel 2018 and IBM SPSS Statistic version 20.0 software (SPSS Inc., Chicago, IL, USA). Significant differences were evaluated using a one-way ANOVA analysis. Tukey's HSD test was selected, and tests were conducted under 95% of confidence interval.

3. Results and Discussion

3.1. LED Light Dose and Temperature Profile

Figure 1 depicts a typical temperature profile recorded upon illumination at three different irradiances (10.4, 28.9 and 36.2 mW·cm^{-2}). An increase in the irradiance caused a rise in the temperature inside the light-box. The highest irradiance resulted in an increase of temperature by 12.8 and 19.67 °C after 20 and 30 min of illumination, respectively. On the other hand, the temperature only increased by 2.34 °C when the lowest irradiance was used. An increase in temperature has also been reported in several studies on photosensitization using LED [32]. Although small fans were installed in the light chamber, this did not prevent the temperature increase during photosensitization.

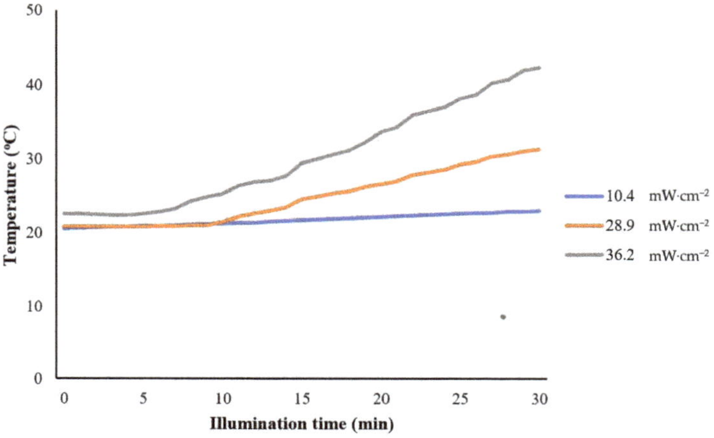

Figure 1. Temperature profile inside the light chamber upon blue LED illumination at three different irradiances (10.4, 28.9 and 36.2 mW·cm^{-2}).

3.2. In Vitro Experiments

3.2.1. Effect of Photosensitization on *A. flavus* Spores

The efficiency of *A. flavus* spores' inactivation by photosensitization was dependent on the light dose applied for the treatment. Whereas, using the same solvent (ethanol or propylene glycol), the tested concentrations of curcumin did not show a significant difference ($p > 0.05$) in their inactivation efficiency. Photosensitization (P^+L^+) significantly ($p < 0.05$) reduced the viability of *A. flavus* spores using light doses above 34.7 J·cm^{-2} (Figure 2). Compared to the negative control (P^oL^o), no significant change was observed on the population of spores that were either treated with the photosensitizer (P^+L^o) or light (P^oL^+) (Figure 2), indicating a synergic effect from the combination of light and curcumin in reducing the multiplication of *A. flavus*. The efficiency of inactivation was not compared between the solvents as different concentrations of curcumin were used for each solvent.

An increase in the light dose to 86.9 J·cm^{-2} enhanced photosensitization efficiency. However, further increase of light dose to 130.3 J·cm^{-2} did not significantly increase ($p > 0.05$) the efficiency (Figure 2). The presence of curcumin in ethanolic solution resulted in the highest inactivation with 1.9 log CFU·mL^{-1} reduction at 104.2 J·cm^{-2} using 25 μM curcumin. However, the same light dose combined with 40-fold higher curcumin concentration (1000 μM prepared in propylene glycol) resulted in 2.3 log CFU·mL^{-1} reduction. Since curcumin is partially soluble in propylene glycol, higher concentrations of the photosensitizer are required to achieve a significant fungal inactivation as previously reported [24]. Treatments with light doses below 34.7 J·cm^{-2} did not show any significant effects on the viability of *A. flavus* spores ($p > 0.05$) compared to the negative control P^oL^o (data not shown). Furthermore, the spores did not exhibit any susceptibility to ethanol or propylene glycol solutions, as evaluated by the inhibition zone through well-diffusion assay, suggesting that the solvents used are non-toxic to the *A. flavus* spores at the concentrations applied in this study.

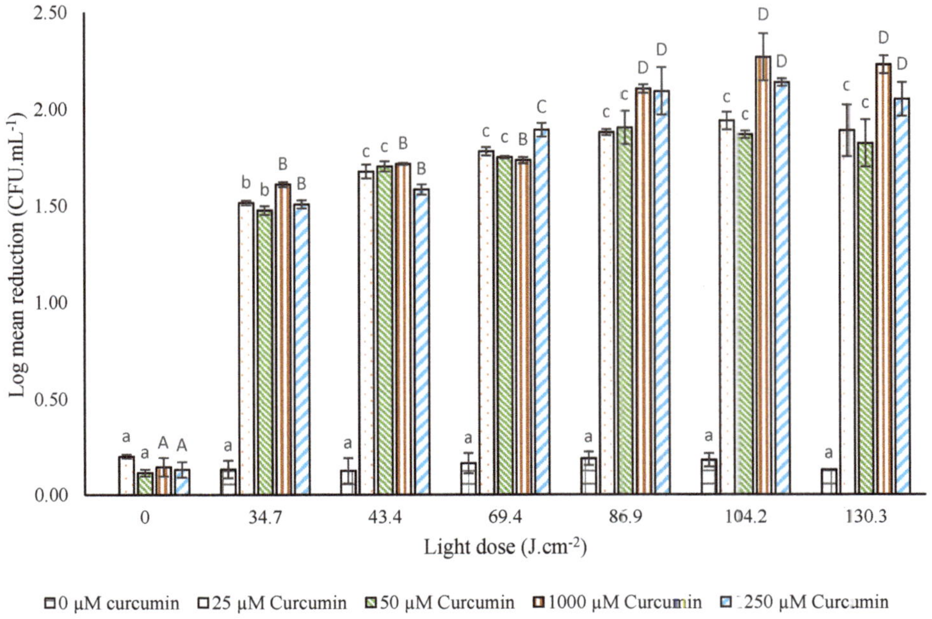

Figure 2. Effect of curcumin-based photosensitization at different curcumin concentrations, prepared in ethanol and propylene glycol, and light doses on reduction of *Aspergillus flavus* spores in vitro. Data are means ± SE (n = 9); means with different lowercase letters indicate significant ($p < 0.05$) differences at different light doses and means with uppercase letters indicate significant ($p < 0.05$) differences at the same light dose.

Although an increase in temperature inside the box was recorded during illumination, the microbial inactivation was achieved as a result of a photodynamic effect of curcumin in the presence of light, but not the heating effect. This was ascertained by submitting the spores to the light alone, which did not show a significant effect on spores' viability even though a temperature as high as 42.27 °C was achieved for 130.4 J·cm^{-2} light dose. *Aspergillus flavus* can grow in temperatures ranging from 12 to 48 °C, although the optimum growth occurs at 37 °C [33]. Moreover, variation of fluid temperature from 15 to 45 °C has been reported to have no significant impact on *A. flavus* viability [23]. Therefore, the temperatures achieved upon LED illumination were not high enough to inactivate the spores.

The phototoxicity of curcumin has been demonstrated against fungi, bacteria and virus [8,9,24]. Although the exact mechanism of microbial inactivation through photosensitization is still not well understood, it has been proposed as a result of ROS formation when the photosensitizer is illuminated at an appropriate wavelength [34,35]. In turn, ROS react with cellular components such as lipids, proteins and nucleic acids, ultimately causing microbial cell death [5]. Results obtained from the present study suggest that 430-nm LEDs are capable of exciting curcumin to the triplet state, and when returning to the ground state, the photosensitizer combines with oxygen resulting in the generation of ROS.

3.2.2. Reactive Oxygen Species (ROS) Generation

To determine ROS formation, 50 μM and 250 μM curcumin prepared in ethanol and propylene glycol solution, respectively, were selected. The former was the optimum working concentration in vitro, whereas the latter, however, was selected since, although the 1000 µM curcumin in propylene glycol resulted in the highest photoinactivation of *A. flavus*, the absorbance was above the maximum detection limit of the spectrophotometer used (absorbance over 2.5). Therefore, a lower concentration of curcumin in propylene glycol was used in the experiment.

The ROS generation was indirectly monitored using DPBF, a reactive oxygen species quencher [36]. The absorption intensity of DPBF at 410 nm provides a qualitative estimation of ROS generated during photosensitization, and its decrease indicates the generation of ROS. Light intensity of 36.2 $mW \cdot cm^{-2}$ led to a higher and more rapid reduction in DPBF absorption intensity, which in turn represents a swift generation of ROS (Figure 3). Additionally, illumination at 36.2 $mW \cdot cm^{-2}$ ($k_{36.2}$ = 2.2277) gave rise to 1.4 times faster ROS generation in ethanolic curcumin solution than, illumination at 28.9 mW/cm^{-2} ($k_{28.9}$ = 1.622). For curcumin dissolved in propylene glycol, the ROS were generated 1.6 times faster at 36.2 $mW \cdot cm^{-2}$, than at 28.9 $mW \cdot cm^{-2}$ ($k_{36.2} = -0.7561$ and $k_{28.9} = -0.4841$, respectively). The DPBF decay in ethanolic curcumin solution followed second-order kinetics (Figure 3c). However, first-order kinetics was observed when using propylene glycol as the solvent (Figure 3d).

The use of ethanolic curcumin solution ensures a generation of a higher amount of ROS than curcumin in propylene glycol, though higher concentration was used in the latter solvent. These results correlate with the observed photoinactivation efficacy of curcumin in ethanol and propylene glycol against *A. flavus* spores. To achieve an effective microbial inactivation, higher curcumin concentrations were required using propylene glycol (>500 µM). Whereas curcumin concentrations as low as 25 µM were effective against the spores using ethanol as solvent. No significant change ($p > 0.05$) was observed in the absorption intensity of DPBF through dark incubation with the photosensitizer.

Curcumin is highly soluble in ethanol but showed partial solubility in propylene glycol. Therefore, it is reasonable to assume that all curcumin dissolved in the ethanolic solution is readily available to absorb light and produce ROS, resulting in a high efficiency at a low concentration. Nevertheless, using propylene glycol, a large amount of curcumin is needed to produce enough quantum yield to photoinactivate the spores, as it was previously reported. Using a propylene glycol solution of curcumin to inactivate several fungi species, including *Aspergillus* spp., Al-Asmari et al. [24] observed that dye concentrations over 600 µM were required to inactivate roughly 90% of fungal spores. However, concentrations between 100 and 400 µM inactivated only up to 60% of the spores. The same authors evaluated the efficacy of curcumin in propylene glycol to inactivate the naturally occurring microflora in date fruits and reported 1400 µM of curcumin as the most effective concentration against various microorganisms [8]. On the other hand, Temba, Fletcher, Fox, Harvey and Sultanbawa [22] observed that 15–50 µM of curcumin dissolved in ethanol were effective in reducing the colony-forming ability of *A. flavus*. A reduction of up to 3 log units could be observed. However, a further increase of the curcumin concentration to 100 µM did not cause any additional inactivation of the spores.

DPBF is the most used quencher to estimate the concentration of ROS in media. The decrease of DPBF intensity is due to its oxidation with ROS generated upon illumination of the photosensitizer solution. On the other hand, incubation in the dark showed no reduction in the DPBF absorption intensity since the generation of ROS to interact with the probe does not occur when the photosensitizer is incubated in the absence of light [37,38]. The results show that ROS are efficiently generated upon the irradiation of curcumin with 430-nm LED, and the higher the concentration of curcumin, the more DPBF is consumed which suggests a high generation of ROS.

Differences in ROS yielded by curcumin upon illumination has also been reported in previous studies [39]. A hypothesis explaining the differences in ROS yield of a photosensitizer based on the solvent used was proposed by Geroge and Kishen [28] who studied the ROS generation by Methylene Blue (MB) dissolved in different solvents such as 70% polyethylene glycol (PEG), 70% glycerol, and a mixture of glycerol: ethanol: water (30:20:50, $v/v/v$). The authors observed that the photobleaching of DPBF was lower in PEG and glycerol than in the formulation containing ethanol. It was suggested that the difference could be due to the type of interaction between the solvent molecules and the excited state of the photosensitizer. Nevertheless, understanding the effect of the solvent on the ROS generation yield by the photosensitizer is a subject that merits a thorough evaluation in future studies.

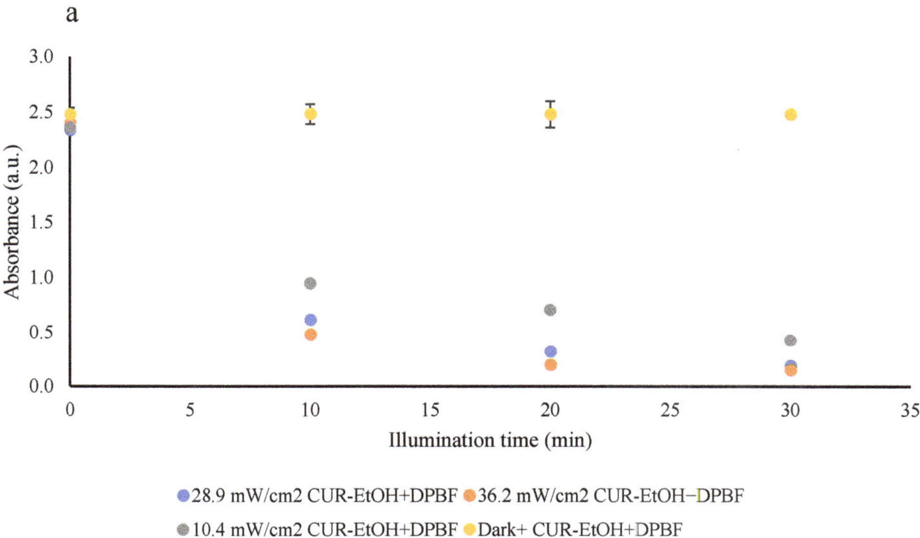

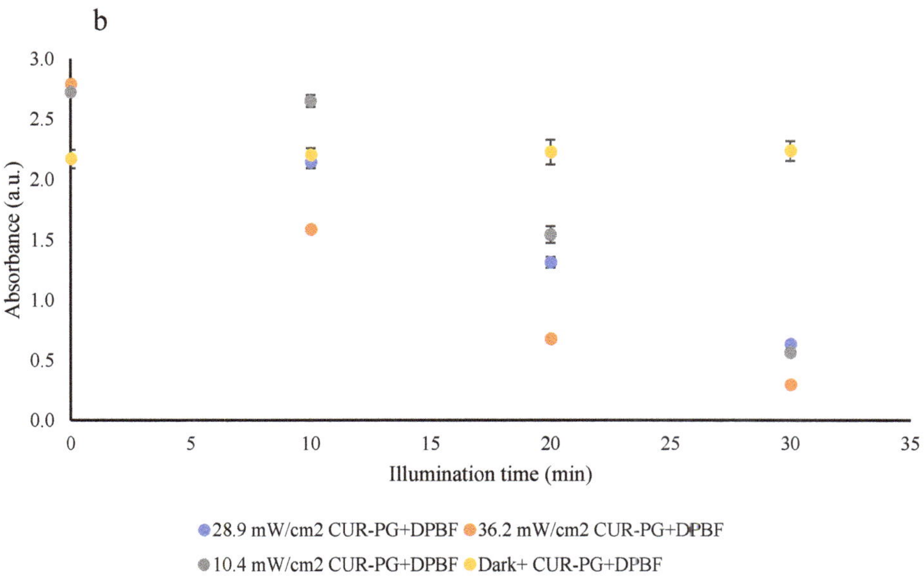

Figure 3. *Cont.*

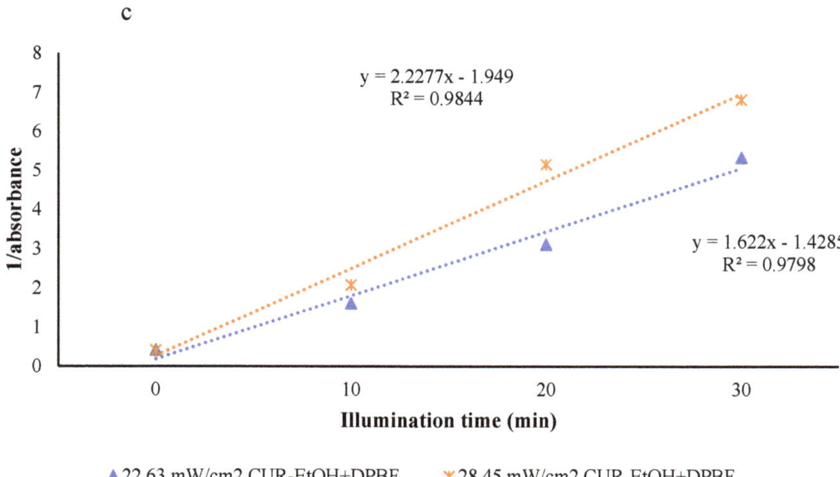

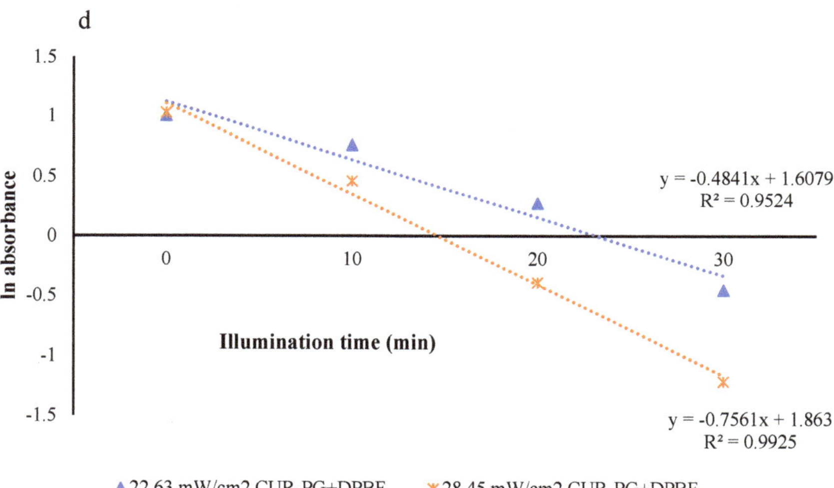

Figure 3. (**a**) Decay of DPBF absorbance upon illumination (EtOH-CUR = curcumin in ethanol); (**b**) Decay of DPBF absorbance upon illumination (CUR-PG = curcumin in 50% propylene glycol); (**c**) Kinetic decay of DPBF absorbance in ethanolic curcumin solution (second-order kinetics); (**d**) Kinetic decay of DPBF absorbance in curcumin-propylene glycol solution (first-order kinetics). Data are means ± SE (n = 3).

3.2.3. Morphological Changes of *A. flavus* Spores

Figure 4 presents representative SEM micrographs of *A. flavus* spores subjected to different treatments. The results show that spores treated with light had lower density and some cell wall deformations. On the other hand, non-treated (control) and curcumin-treated cells showed higher density and hyphal filament with no apparent deformations with the normal globular shape. In contrast, photosensitized spores showed the least cell density with some distortions in cellular shape, where the extent of damage was enough to cause cell death. Some cells in the photosensitization treatment still showed normal shape which substantiates the fact that some of the cells were still viable in vitro. The reduction in cell density due to photosensitization has been previously reported in *Candida* spp. [40,41].

The cell wall is dynamic and undergoes changes in response to oxidative stress. ROS generated during photosensitization react with macromolecules such as proteins, polysaccharides, lipids and nucleic acids causing molecular changes that can negatively affect cellular physiology and morphology [42]. Deformations of the cell wall may result in intracellular content leakage, causing the cell shrinking and ultimately cell death [43] which has been reported for several microorganisms such as *Trichophyton rubrum* [29], *Escherichia coli* [43], and *Candida albicans* [44].

It was observed that light treatment alone was able to cause deformations of the cell wall in some spores, which is in agreement with previous reports [29]. Nevertheless, Smijs et al. [29] reported that changes caused by light treatment on *T. rubrum* spores were reversible, and the spores exhibited normal growth after eight days of incubation. This could explain the observed normal growth of light treated spores.

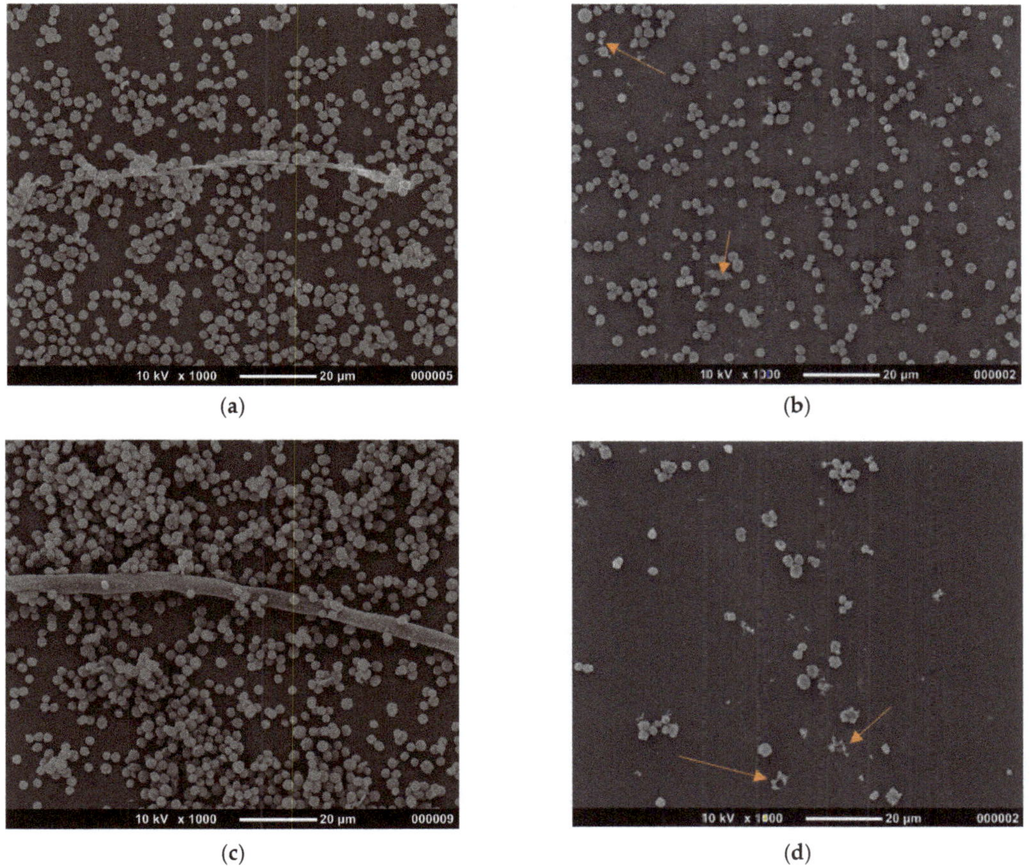

Figure 4. Scanning electron micrographs of *Aspergillus flavus* spores after different treatments (×1000; scale bar = 20 µm). (**a**) P^oL^o = control; (**b**) P^oL^+ = light; (**c**) P^+L^o = photosensitizer; (**d**) P^+L^+ = photosensitization. Arrows in the SEM micrographs indicate cell wall deformation.

3.3. *In Vivo Experiment*

3.3.1. Photodecontamination of *A. flavus* on Maize

The combination of 1000 µM of curcumin in propylene glycol and 102.4 J·cm^{-2} was selected according to the in vitro results which had showed to inactivate *A. flavus* spores (Section 3.2). The average initial spore's load in popcorn, yellow and white maize were 5.78,

5.44 and 5.61 log CFU·g^{-1}, respectively. The viability of spores was not affected ($p > 0.05$) by curcumin or light treatment. Moreover, up to 0.39 ± 0.06 log CFU·g^{-1} reduction was obtained through 25% propylene glycol application on maize kernels (Figure 5), which corresponds to the concentration of the solvent in 1000 µM of curcumin. On the other hand, *A. flavus* colonies were not observed in the experiments with photosensitized maize kernel extract, suggesting that photosensitization completely removed the spores from the kernel surface, which was in contrast to the in vitro results, where only 2.3 log CFU·mL^{-1} reduction was achieved. The immersion of kernel in curcumin solution for 5 min before illumination might have washed the spores on its surface, reducing the microbial load before illumination. Consequently, the initial concentration of spores on the maize surface may have been lower than that used in vitro. Complete decontamination of food through curcumin-mediated photosensitization has also been reported in oysters inoculated with *Vibrio parahaemolyticus* [45].

Contrary to our results, several studies have reported lower photoinactivation efficiency on food surface than in vitro [46,47], which have been attributed to the irregularity and different light-reflecting properties of the food under illumination [48]. In a study reported by Temba et al. [22], nearly 2 log CFU·g^{-1} reduction was achieved on maize kernel and flour through the combination of curcumin and light, against 3 log CFU·g^{-1} reduction achieved in vitro. This divergence of the results by Temba et al. [22] and those reported in the present study might be related to the use of different light doses, 60 J·cm^{-2} (Temba et al.) vs. 102.4 J·cm^{-2} (present study). In addition, the illumination source used by Temba and co-workers [22] was unidimensional, and though the kernels were constantly turned over during illumination, the light may have been distributed irregularly, reducing the effectiveness of the treatment. Nevertheless, these are mere assumptions, and further studies are necessary to explore the effect of the maize matrix on the efficiency of photosensitization.

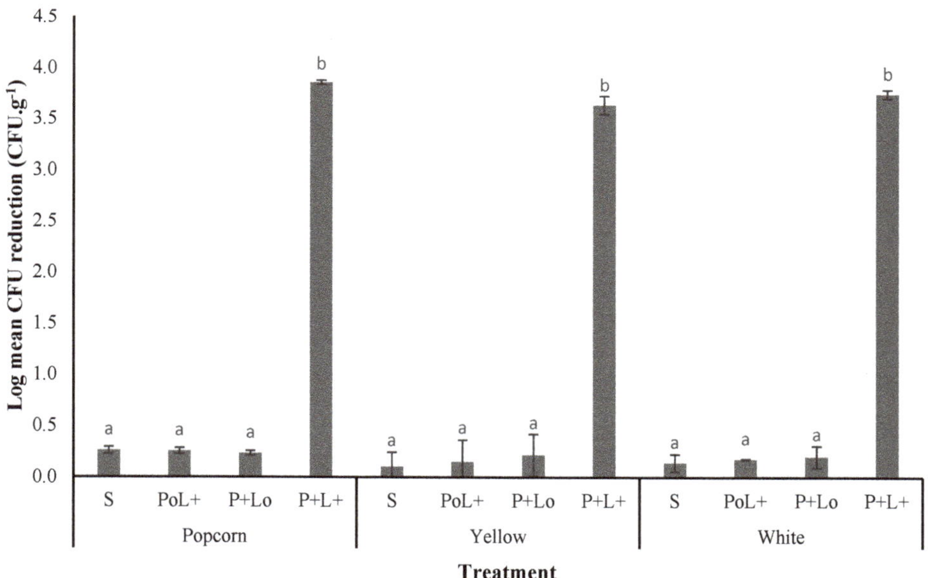

Figure 5. Photodecontamination of *Aspergillus flavus* spores on three maize varieties (popcorn, yellow and white). S = solvent (propylene glycol); P^oL^+ = light; P^+L^o = photosensitizer; P^+L^+ = photosensitization. Data are means ± SE (n = 3); means with different letters in the same column are significantly different ($p < 0.05$).

3.3.2. Effect of Photosensitization on AFB1 Production

The initial concentration of AFB1 was below the limit of quantification (LOQ = 0.01 $\mu g \cdot L^{-1}$) in all treatments and maize kernels. After seven days of incubation, AFB1 was still undetectable (below the LOQ) both in photosensitized and curcumin-treated maize kernels, whereas the control and the light treated ones showed AFB1 accumulation over incubation time. Furthermore, there were no differences ($p > 0.05$) in AFB1 accumulation amongst the investigated maize varieties (Figure 6).

From seven to fourteen days of incubation, a significant ($p < 0.05$) increase in AFB1 concentration was observed in all treatments, except for the photosensitized kernels where AFB1 was still undetectable. The curcumin-treated kernels showed an increase by over 60-fold after 14 days of incubation, reaching more than 63 $mg \cdot kg^{-1}$ AFB1 in white, yellow and popcorn maize. On the other hand, the light-treated white kernels showed a 18-fold increase in AFB1 concentration after 14 days. Yellow and popcorn varieties showed a significant ($p < 0.05$) increase to over 335.08 $mg \cdot kg^{-1}$ AFB1. The control was not different ($p > 0.05$) from the light treatment in all three varieties.

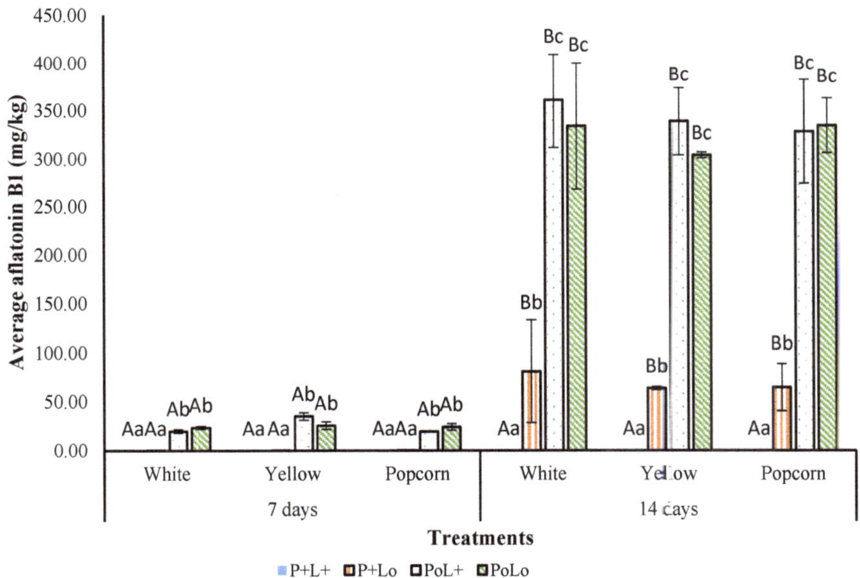

Figure 6. Concentration of aflatoxin B1 on maize kernel after photosensitization (P^+L^+), photosensitizer (P^+L^+), light (P^oL^+) treatments and negative control (P^oL^o). Data are means ± SE ($n = 3$); means followed by different letters (lowercase letter for treatments in the same variety and uppercase letter for different periods in the same variety) are significantly different ($p < 0.05$).

Findings from this study substantiate the potential of photosensitization as an effective clean green technology to decontaminate food and reduce mycotoxin exposure to consumers, which is in agreement with previous studies [4,49]. Curcumin-mediated photosensitization has been reported to reduce the accumulation of AFB1 by 73% on maize kernel after 10 days of incubation at moist conditions [23]. In another study, the combination of blue light and toluidine blue-O, a synthetic photosensitizer, reduced the formation of alternariol by *Alternaria alternata* by over 90% for isolates from squash, tomato, pepper and cucumber [49].

Our results on AFB1 formation corroborates with the visual observations of mould growth on maize kernel. In photosensitized kernels, the colony-forming ability of *A. flavus* was completely suppressed, and no fungal growth was visually observed on the maize samples even after 14 days of incubation at moist conditions. On the other hand, AFB1 was not detectable in curcumin-treated kernels after seven days of incubation, which is consistent with no fungal growth observation in the same period. It could be suggested that propylene glycol in curcumin solution is acting as a coating layer on the maize surface, encapsulating curcumin and releasing it throughout the seven days of incubation, which increases the contact time between curcumin and spores, reducing their viability. The antimicrobial properties of curcumin against different fungal pathogens, including *A. flavus*, have been documented in previous reports [50–52]. Additionally, soaking the kernels in curcumin solution might have partially washed the spores from the kernel surface reducing their concentration and delaying germination during the incubation period.

3.3.3. Effect of Photosensitization on Maize Colour

To determine whether photosensitization had any effect on the colour of maize, the kernels were analysed immediately after treatment. Although photosensitization did not have a significant effect ($p > 0.05$) on the colour parameters, i.e., lightness index (L*), greenness index (a*) and yellowness index (b*), and chroma, the mean CIELAB Delta-E (ΔE) value was over 3.0 (Table 1), implying that the difference between the control and photosensitized maize kernels were perceptive to an untrained eye, as indicated by Pathare et al. [53]. Photosensitization of the kernels faded the curcumin owing to its light susceptibility, especially in neutral-basic conditions [54]. Similarly, a loss in yellow-orange colour of riboflavin was observed after photosensitization of smoked salmon, where riboflavin was used as a photosensitizer [55].

Table 1. Colour parameters in maize kernels submitted to different treatments.

Treatment	L*	a*	b*	C	ΔE
P^oL^o white	81.9 ± 0.5 a	-0.7 ± 0.3 a	29.4 ± 0.9 b	29.4 ± 0.9 b	-
P^+L^o white	81.7 ± 0.8 a	-6.5 ± 0.4 b	50.6 ± 1.8 a	51.0 ± 1.7 a	21.9 a
P^+L^+ white	79.8 ± 0.5 a	0.1 ± 0.4 a	26.8 ± 0.7 b	26.8 ± 0.7 b	3.5 b
P^oL^o yellow	74.8 ± 0.5 a	6.8 ± 0.7 ab	39.8 ± 2.5 b	40.3 ± 2.5 b	-
P^+L^o yellow	77.6 ± 1.3 a	4.4 ± 0.9 b	54.1 ± 1.9 a	54.2 ± 1.9 a	14.7 a
P^+L^+ yellow	77.2 ± 0.5 a	7.5 ± 0.6 a	37.6 ± 2.4 b	38.3 ± 2.3 b	3.3 b
P^oL^o popcorn	74.1 ± 0.6 a	9.8 ± 0.9 a	28.5 ± 1.3 a	30.2 ± 1.4 b	-
P^+L^o popcorn	77.4 ± 0.9 a	5.8 ± 1.0 b	34.9 ± 1.8 a	35.3 ± 1.7 a	8.7 a
P^+L^+ popcorn	76.0 ± 0.7 a	8.7 ± 0.7 a	25.4 ± 1.9 a	26.9 ± 1.8 b	3.8 b

P^oL^o = control; P^+L^o = curcumin treatment; P^+L^+ = photosensitization; C = chroma; data are means ± SE ($n = 3$); data with different letters in the same column and maize variety are significantly different ($p < 0.05$). The comparison of means was performed within the same variety for different treatments, as the aim was to assess if different treatments result (or not) in colour changes on the surface of the maize kernel.

The obtained extracts from photosensitized maize kernels showed a peak in the region of curcumin maxima absorption, i.e., 410–430 nm (Figure 7). The existence of a peak indicates that 5-min soaking of maize resulted in curcumin absorption to the inner parts of the kernel, which was unlikely photobleached upon illumination. The blue LED light (400–470 nm) is a superficial decontamination technique, with a depth penetration below 1 mm [56]. Therefore, the LED light did not penetrate the maize matrix, and only the curcumin existing on the kernel surface was degraded or excited for ROS production.

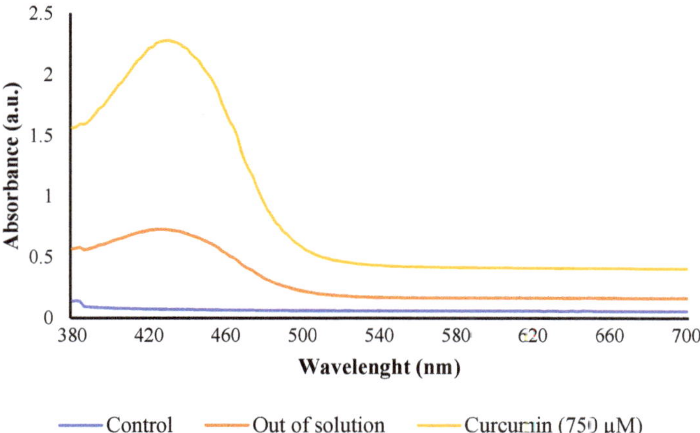

Figure 7. Maize extract absorption spectra after soaking in curcumin solution and illumination with blue light at 104.2 J·cm^{-2}. Control maize kernels dipped in the solvent before illumination (propylene glycol); out of solution maize kernels dipped in 1000 μM of curcumin in propylene glycol before illumination.

3.4. Effect of Blue Light on Carotenoid Content in Maize Kernels

The illumination of maize kernels was conducted at 130.3 J·cm^{-2} (i.e., 36.2 mW·cm^{-2}, 30 min), the most extreme studied conditions for *A. flavus* photoinactivation. The white variety presented the lowest content of total carotenoids among the three varieties studied, with an initial content of 1.74 μg·g^{-1} dry weight (DW), followed by the yellow variety with the medium content of 12.90 μg·g^{-1} DW. The initial content of carotenoids in popcorn maize was about 19-fold and 2.6-fold of that of white and yellow maize, respectively, with a total content of 33.44 μg·g^{-1} DW (Table 2). Zeaxanthin was the predominant carotenoid compound in all varieties, and its highest concentration was found in the popcorn accession of up to 24.04 μg·g^{-1} DW.

Table 2. Carotenoid content in the studied maize varieties (white, yellow and popcorn) before and after illumination using LED blue light at 130.3 J·cm^{-2}.

Treatment	Individual Carotenoids	Carotenoid Content (μg.g^{-1} DW)		
		White	Yellow	Popcorn
Control	Lutein	0.22 ± 0.02 ^{a}	5.69 ± 0.19 ^{b}	5.67 ± 0.07 ^{b}
	Zeaxanthin	1.54 ± 0.01 ^{a}	7.20 ± 0.14 ^{b}	24.04 ± 0.26 ^{c}
	β-cryptoxanthin	n.d.	n.d.	1.86 ± 0.07 ^{b}
	β-carotene	n.d.	n.d.	1.84 ± 0.09 ^{b}
	Total	1.74 ± 0.00 ^{Aa}	12.90 ± 0.32 ^{Ab}	33.44 ± 0.48 ^{Ac}
Illumination	Lutein	0.17 ± 0.01 ^{a}	6.07 ± 0.48 ^{b}	5.76 ± 0.24 ^{b}
	Zeaxanthin	1.45 ± 0.02 ^{a}	7.29 ± 0.42 ^{b}	24.67 ± 0.70 ^{c}
	β-cryptoxanthin	n.d.	n.d.	2.29 ± 0.07 ^{a}
	β-carotene	n.d.	n.d.	3.30 ± 0.076 ^{b}
	Total	1.62 ± 0.01 ^{Aa}	13.36 ± 0.91 ^{Ab}	36.01 ± 1.05 ^{Ac}

Data are means ± SE (n = 3); data with different letters in the same row are significantly different ($p < 0.05$); n.d.: not detectable.

The use of LED light did not affect ($p > 0.05$) the content of carotenoids in the investigated maize varieties. Light exposure might cause carotenoid degradation resulting in the formation of carotenoid radical cations [57]. However, the wavelengths mostly responsible for carotenoid degradation are within the UV range [58]. The light used in this study was within the visible range, and the average illuminance for a light dose of 130.3 J·cm^{-2} was

2843 ± 16.15 lux. Therefore, although effective on reducing the viability of *A. flavus* spores, the irradiance and illumination time used were not able to cause a significant degradation or isomerization of carotenoids in the three maize varieties.

Similarly, previous studies in fruits have also reported a non-significant effect of LED illumination on the nutritional content, although an increase in some compounds was observed. Kim et al. [59] evaluated the effect of LED illumination on antioxidant capacity and the content of beta-carotene, flavonoids, lycopene, and ascorbic acid in fresh-cut mango. It was shown that illumination did not affect the antioxidant capacity of fruits and the content of β-carotene, ascorbic acid, and lycopene, while the content of flavonoids increased 1.9 times. More than 95% of β-carotene could be retained in tomato juice by using a constant illuminance of 2476 lux for 24 h at 4 ^{0}C [60]. These results substantiate the potential of using LEDs in photosensitization to reduce the microbial load on the maize surface without affecting carotenoids.

Previous studies have reported a reduction in mycotoxin accumulation in the presence of carotenoids [15,16]. However, no significant differences were observed in AFB1 accumulation between white maize having a relatively low carotenoid content of 1.74 $\mu g \cdot g^{-1}$ DW and popcorn maize with a relatively high content of 33.44 $\mu g \cdot g^{-1}$ DW. In addition, the inhibition of mycotoxin accumulation by carotenoids is strain-dependent [14,15]. Furthermore, it has also been reported that the reduction in mycotoxin production depends on the individual carotenoids present, particularly β-carotene and β-cryptoxanthin have a significant effect on the aflatoxin synthesis [16]. In the present study, only the popcorn maize had detectable levels of β-cryptoxanthin and β-carotene, while zeaxanthin was the predominant carotenoid in all varieties. Furthermore, technical challenges such as developing the most efficient procedure to disperse the curcumin solution, need to be addressed, when upscaling this new green preservation technique to farm/industry-scale.

4. Conclusions

Curcumin in combination with 430 nm LEDs reduced the viability of *A. flavus* spores in vitro and on the surface of maize kernels, resulting in an inhibition of the synthesis of aflatoxin B1. At a given curcumin concentration, the inactivation efficiency was light-dose dependent. The highest fungal inactivation was achieved using 1000 μM curcumin and a light dose of 81.5 J/cm^2. The use of photosensitization did not cause significant changes in the carotenoid content as well as the colour attributes of the maize kernels. The use of novel and green technologies such as photosensitization could be a valuable alternative to tackle the challenges posed by aflatoxins and their toxicity for humans. The reduction of fungal contamination on maize kernels could present a positive step towards the mitigation of the current level of postharvest losses in maize, and also enhance the nutritional value and quality of this important crop. Optimization of the curcumin concentration to minimize its effect on the sensory quality of food products is crucial and will benefit not only the food industry but also consumers.

Author Contributions: Conceptualization, Y.S., M.S.D. and R.N.; methodology, Y.S., M.S.D., R.N., A.D.T.P., H.T.H., M.C., M.E.N. and T.J.O.; formal analysis, R.N.; investigation, R.N. and M.S.D.; data curation, and writing—original draft preparation, R.N.; writing—review and editing, Y.S., M.S.D., A.D.T.P., H.T.H., M.E.N. and R.N.; funding acquisition and supervision, Y.S. All authors have read and agreed to the published version of the manuscript.

Funding: This research was supported by the Australian Government through an Australia Awards Scholarship (AUSAID-DFAT) for Master of Science studies, the University of Queensland, Horticulture Innovation Australia Ltd. (Naturally Nutritious project HN15001) and the Department of Agriculture and Fisheries.

Institutional Review Board Statement: Not applicable.

Informed Consent Statement: Not applicable.

Data Availability Statement: The data presented in this study are available on request from the corresponding author.

Acknowledgments: The authors would like to thank Andrew Cusack and Margaret Currie for their technical assistance.

Conflicts of Interest: The authors declare no conflict of interest.

References

1. Fink-Gremmels, J.; van der Merwe, D. Mycotoxins in the food chain: Contamination of foods of animal origin. In *Chemical Hazards in Foods of Animal Origin*; Smulders, F.J.M., Rietjens, I.M.C.M., Rose, M., Eds.; Wageningen Academic Publishers: Wageningen, The Netherlands, 2019; Volume 7, pp. 1190–1198.
2. Sarma, U.P.; Bhetaria, P.J.; Devi, P.; Varma, A. Aflatoxins: Implications on health. *Indian J. Clin. Biochem.* **2017**, *32*, 124–133. [CrossRef]
3. Al-Jaal, B.; Salama, S.; Al-Qasmi, N.; Jaganjac, M. Mycotoxin contamination of food and feed in the Gulf Cooperation Council countries and its detection. *Toxicon* **2019**, *171*, 43–50. [CrossRef] [PubMed]
4. Glueck, M.; Schamberger, B.; Eckl, P.; Plaetzer, K. New horizons in microbiological food safety: Photodynamic Decontamination based on a curcumin derivative. *Photochem. Photobiol. Sci.* **2017**, *16*, 1784–1791. [CrossRef] [PubMed]
5. Hu, J.; Lin, S.; Tan, B.; Hamzah, S.; Lin, Y.; Kong, Z.; Zhang, Y.; Zheng, B.; Zeng, S. Photodynamic inactivation of *Burkholderia cepacia* by curcumin in combination with EDTA. *Food Res. Int.* **2018**, *111*, 265–271. [CrossRef] [PubMed]
6. Ghate, V.S.; Zhou, W.; Yuk, H.G. Perspectives and trends in the application of photodynamic inactivation for microbiological food safety. *Compr. Rev. Food Sci. Food Saf.* **2019**, *18*, 402–424. [CrossRef] [PubMed]
7. de Oliveira, E.; Tosati, J.; Tikekar, R.; Monteiro, A.; Nitin, N. Antimicrobial activity of curcumin in combination with light against *Escherichia coli* O157: H7 and *Listeria innocua*: Applications for fresh produce sanitation. *Postharvest Biol. Technol.* **2018**, *137*, 86–94. [CrossRef]
8. Al-Asmari, F.; Mereddy, R.; Sultanbawa, Y. The effect of photosensitization mediated by curcumin on storage life of fresh date (*Phoenix dactylifera* L.) fruit. *Food Control.* **2018**, *93*, 305–309. [CrossRef]
9. Liu, F.; Li, Z.; Cao, B.; Wu, J.; Wang, Y.; Xue, Y.; Xu, J.; Xue, C.; Tang, Q.J. The effect of a novel photodynamic activation method mediated by curcumin on oyster shelf life and quality. *Food Res. Int.* **2016**, *87*, 204–210. [CrossRef]
10. Priyadarsini, K.I. Photophysics, photochemistry and photobiology of curcumin: Studies from organic solutions, bio-mimetics and living cells. *J. Photochem. Photobiol. C* **2009**, *10*, 81–95. [CrossRef]
11. Thrane, J.-E.; Kyle, M.; Striebel, M.; Haande, S.; Grung, M.; Rohrlack, T.; Andersen, T. Spectrophotometric analysis of pigments: A critical assessment of a high-throughput method for analysis of algal pigment mixtures by spectral deconvolution. *PLoS ONE* **2015**, *10*, e0137645. [CrossRef]
12. Lichtenthaler, H.K.; Buschmann, C. Chlorophylls and carotenoids: Measurement and characterization by UV-VIS spectroscopy. *Curr. Protoc. Food Anal. Chem.* **2001**, *1*, F4. 3.1–F4. 3.8. [CrossRef]
13. Meléndez-Martínez, A.J.; Stinco, C.M.; Mapelli-Brahm, P. Skin carotenoids in public health and nutricosmetics: The emerging roles and applications of the UV radiation-absorbing colourless carotenoids phytoene and phytofluene. *Nutrients* **2019**, *11*, 1093. [CrossRef] [PubMed]
14. Norton, R.A. Effect of carotenoids on aflatoxin B1 synthesis by *Aspergillus flavus*. *Phytopathology* **1997**, *87*, 814–821. [CrossRef] [PubMed]
15. Wicklow, D.; Norton, R.; McAlpin, C. β-Carotene inhibition of aflatoxin biosynthesis among *Aspergillus flavus* genotypes from Illinois corn. *Mycoscience* **1998**, *39*, 167–172. [CrossRef]
16. Suwarno, W.; Hannok, P.; Palacios-Rojas, N.; Windham, G.; Crossa, J.; Pixley, K. Provitamin A carotenoids in grain reduce aflatoxin contamination of maize while combating vitamin A deficiency. *Front. Plant Sci.* **2019**, *10*, 30. [CrossRef]
17. Aman, R.; Schieber, A.; Carle, R. Effects of heating and illumination on trans– cis isomerization and degradation of β-carotene and lutein in isolated spinach chloroplasts. *J. Agric. Food Chem.* **2005**, *53*, 9512–9518. [CrossRef] [PubMed]
18. Ramel, F.; Birtic, S.; Cuiné, S.; Triantaphylidès, C.; Ravanat, J.-L.; Havaux, M. Chemical quenching of singlet oxygen by carotenoids in plants. *Plant Physiol.* **2012**, *158*, 1267–1278. [CrossRef]
19. Jung, M.; Min, D. Effects of quenching mechanisms of carotenoids on the photosensitized oxidation of soybean oil. *J. Am. Oil Chem. Soc.* **1991**, *68*, 653–658. [CrossRef]
20. Dutta, D.; Chaudhuri, U.; Chakraborty, R. Structure, health benefits, antioxidant property and processing and storage of carotenoids. *Afr. J. Biotechnol.* **2005**, *4*, 1510–1520. [CrossRef]
21. Oshima, S.; Ojima, F.; Sakamoto, H.; Ishiguro, Y.; Terao, J. Inhibitory effect of β-carotene and astaxanthin on photosensitized oxidation of phospholipid bilayers. *J. Nutr. Sci. Vitaminol.* **1993**, *39*, 607–615. [CrossRef]
22. Temba, B.A.; Fletcher, M.T.; Fox, G.P.; Harvey, J.J.W.; Sultanbawa, Y. Inactivation of *Aspergillus flavus* spores by curcumin-mediated photosensitization. *Food Control* **2016**, *59*, 708–713. [CrossRef]
23. Temba, B.A.; Fletcher, M.T.; Fox, G.P.; Harvey, J.; Okoth, S.A.; Sultanbawa, Y. Curcumin-based photosensitization inactivates *Aspergillus flavus* and reduces aflatoxin B1 in maize kernels. *Food Microbiol.* **2019**, *82*, 82–88. [CrossRef] [PubMed]

24. Al-Asmari, F.; Mereddy, R.; Sultanbawa, Y. A novel photosensitization treatment for the inactivation of fungal spores and cells mediated by curcumin. *J. Photochem. Photobiol. B* **2017**, *173*, 301–306. [CrossRef] [PubMed]
25. Maclean, M.; MacGregor, S.J.; Anderson, J.G.; Woolsey, G. Inactivation of bacterial pathogens following exposure to light from a 405-nanometer light-emitting diode array. *Appl. Environ. Microbiol.* **2009**, *75*, 1932–1937. [CrossRef] [PubMed]
26. Gonelimali, F.D.; Lin, J.; Miao, W.; Xuan, J.; Charles, F.; Chen, M.; Hatab, S.R. Antimicrobial properties and mechanism of action of some plant extracts against food pathogens and spoilage microorganisms. *Front. Microbiol.* **2018**, *9*, 1639. [CrossRef]
27. Rout, B.; Liu, C.-H.; Wu, W.-C. Photosensitizer in lipid nanoparticle: A nano-scaled approach to antibacterial function. *Sci. Rep.* **2017**, *7*, 1–14. [CrossRef]
28. George, S.; Kishen, A. Photophysical, photochemical, and photobiological characterization of methylene blue formulations for light-activated root canal disinfection. *J. Biomed. Opt.* **2007**, *12*, 034029. [CrossRef]
29. Smijs, T.; Mulder, A.; Pavel, S.; Onderwater, J.; Koerten, H.; Bouwstra, J. Morphological changes of the dermatophyte *Trichophyton rubrum* after photodynamic treatment: A scanning electron microscopy study. *Med. Mycol.* **2008**, *46*, 315–325. [CrossRef]
30. Sulyok, M.; Berthiller, F.; Krska, R.; Schuhmacher, R. Development and validation of a liquid chromatography/tandem mass spectrometric method for the determination of 39 mycotoxins in wheat and maize. *Rapid Commun. Mass Spectrom.* **2006**, *20*, 2649–2659. [CrossRef]
31. Saini, R.K.; Keum, Y.-S. Carotenoid extraction methods: A review of recent developments. *Food Chem.* **2018**, *240*, 90–103. [CrossRef]
32. Ghate, V.; Kumar, A.; Kim, M.-J.; Bang, W.-S.; Zhou, W.; Yuk, H.-G. Effect of 460 nm light emitting diode illumination on survival of *Salmonella* spp. on fresh-cut pineapples at different irradiances and temperatures. *J. Food Eng.* **2017**, *196*, 130–138. [CrossRef]
33. Yu, J.; Fedorova, N.D.; Montalbano, B.G.; Bhatnagar, D.; Cleveland, T.E.; Bennett, J.W.; Nierman, W.C. Tight control of mycotoxin biosynthesis gene expression in *Aspergillus flavus* by temperature as revealed by RNA-Seq. *FEMS Microbiol. Lett.* **2011**, *322*, 145–149. [CrossRef]
34. Ghorbani, J.; Rahban, D.; Aghamiri, S.; Teymouri, A.; Bahador, A. Photosensitizers in antibacterial photodynamic therapy: An overview. *Laser Ther.* **2018**, *27*, 293–302. [CrossRef]
35. Seidi Damyeh, M.; Mereddy, R.; Netzel, M.E.; Sultanbawa, Y. An insight into curcumin-based photosensitization as a promising and green food preservation technology. *Compr. Rev. Food Sci. Food Saf.* **2020**, *19*, 1727–1759. [CrossRef] [PubMed]
36. Nsubuga, A.; Mandl, G.A.; Capobianco, J.A. Investigating the reactive oxygen species production of Rose Bengal and Merocyanine 540-loaded radioluminescent nanoparticles. *Nanoscale Adv.* **2021**, *3*, 1375–1381. [CrossRef]
37. Diaz-Diestra, D.; Beltran-Huarac, J.; Bracho-Rincon, D.P.; González-Feliciano, J.A.; González, C.I.; Weiner, B.R.; Morell, G. Biocompatible ZnS: Mn quantum dots for reactive oxygen generation and detection in aqueous media. *J. Nanopart. Res.* **2015**, *17*, 1–14. [CrossRef] [PubMed]
38. Hong, S.H.; Koo, M.-A.; Lee, M.H.; Seon, G.M.; Park, Y.J.; Jeong, H.; Kim, D.; Park, J.-C. An effective method to generate controllable levels of ROS for the enhancement of HUVEC proliferation using a chlorin e6-immobilized PET film as a photo-functional biomaterial. *Regen. Biomater.* **2021**, *8*, rbab005. [CrossRef] [PubMed]
39. Chignell, C.; Bilskj, P.; Reszka, K.; Motten, A.; Sik, R.; Dahl, T. Spectral and photochemical properties of curcumin. *Photochem. Photobiol.* **1994**, *59*, 295–302. [CrossRef]
40. Ahmed, H.; Mekawey, A.; Morsy, M. Cellular structural changes in two candida species caused by photodynamic therapy. *Photodiagn. Photodyn. Ther.* **2016**, *4*, 39–54.
41. Sakita, K.M.; Conrado, P.C.; Faria, D.R.; Arita, G.S.; Capoci, I.R.; Rodrigues-Vendramini, F.A.; Pieralisi, N.; Cesar, G.B.; Gonçalves, R.S.; Caetano, W. Copolymeric micelles as efficient inert nanocarrier for hypericin in the photodynamic inactivation of Candida species. *Future Microbiol.* **2019**, *14*, 519–531. [CrossRef] [PubMed]
42. Baptista, A.; Sabino, C.; Núñez, S.; Miyakawa, W.; Martin, A.; Ribeiro, M. Photodynamic damage predominates on different targets depending on cell growth phase of *Candida albicans*. *J. Photochem. Photobiol. B* **2017**, *177*, 76–84. [CrossRef]
43. Pudziuvyte, B.; Bakiene, E.; Bonnett, R.; Shatunov, P.; Magaraggia, M.; Jori, G. Alterations of *Escherichia coli* envelope as a consequence of photosensitization with tetrakis (N-ethylpyridinium-4-yl) porphyrin tetratosylate. *Photochem. Photobiol. Sci.* **2011**, *10*, 1046–1055. [CrossRef]
44. Dong, S.; Shi, H.; Zhang, X.; Chen, X.; Cao, D.; Mao, C.; Gao, X.; Wang, L. Difunctional bacteriophage conjugated with photosensitizers for Candida albicans-targeting photodynamic inactivation. *Int. J. Nanomed.* **2018**, *13*, 2199. [CrossRef] [PubMed]
45. Wu, J.; Mou, H.; Xue, C.; Leung, A.; Xu, C.; Tang, Q.-J. Photodynamic effect of curcumin on *Vibrio parahaemolyticus*. *Photodiagn. Photodyn. Ther.* **2016**, *15*, 34–39. [CrossRef] [PubMed]
46. Luksiene, Z.; Buchovec, I.; Paskeviciute, E. Inactivation of food pathogen *Bacillus cereus* by photosensitization in vitro and on the surface of packaging material. *J. Appl. Microbiol.* **2009**, *107*, 2037–2046. [CrossRef] [PubMed]
47. Tortik, N.; Spaeth, A.; Plaetzer, K. Photodynamic decontamination of foodstuff from Staphylococcus aureus based on novel formulations of curcumin. *Photochem. Photobiol. Sci.* **2014**, *13*, 1402–1409. [CrossRef]
48. Luksiene, Z.; Paskeviciute, E. Novel approach to the microbial decontamination of strawberries: Chlorophyllin-based photosensitization. *J. Appl. Microbiol.* **2011**, *110*, 1274–1283. [CrossRef]
49. Alananbeh, K.; BouQellah, N.; Al Harbi, M.; Ouf, S. The Efficacy of Photosensitizers on Mycelium Growth, Mycotoxin and Enzyme Activity of *Alternaria* spp. *Jordan J. Biol. Sci.* **2018**, *11*, 43–50.

50. Shlar, I.; Droby, S.; Choudhary, R.; Rodov, V. The mode of antimicrobial action of curcumin depends on the delivery system: Monolithic nanoparticles vs. supramolecular inclusion complex. *RSC Adv.* **2017**, *7*, 42559–42569. [CrossRef]
51. Gitika, A.; Mishra, R.; Panda, S.K.; Mishra, C.; Sahoo, P.R. Evaluation of antifungal activity of Curcumin against *Aspergillus flavus*. *Int. J. Curr. Microbiol. Appl. Sci.* **2019**, *8*, 2323–2329. [CrossRef]
52. Pasrija, R.; Kundu, D. Interaction of curcumin with azoles and polyenes against aspergillus infections. *Res. J. Life Sci. Bioinform. Pharm. Chem. Sci.* **2018**, *4*, 271–279.
53. Pathare, P.B.; Opara, U.L.; Al-Said, F.A.-J. Colour measurement and analysis in fresh and processed foods: A review. *Food Bioproc. Technol.* **2013**, *6*, 36–60. [CrossRef]
54. Mondal, S.; Ghosh, S.; Moulik, S.P. Stability of curcumin in different solvent and solution media: UV–visible and steady-state fluorescence spectral study. *J. Photochem. Photobiol. B* **2016**, *158*, 212–218. [CrossRef] [PubMed]
55. Josewin, S.; Ghate, V.; Kim, M.-J.; Yuk, H.-G. Antibacterial effect of 460 nm light-emitting diode in combination with riboflavin against *Listeria monocytogenes* on smoked salmon. *Food Control* **2018**, *84*, 354–361. [CrossRef]
56. Ash, C.; Dubec, M.; Donne, K.; Bashford, T. Effect of wavelength and beam width on penetration in light-tissue interaction using computational methods. *Lasers Med. Sci.* **2017**, *32*, 1909–1918. [CrossRef]
57. da Silva, M.; Paese, K.; Guterres, S.; Pohlmann, A.; Rutz, J.; Cantillano, R.; Nora, L.; de Oliveira Rios, A. Thermal and ultraviolet–visible light stability kinetics of co-nanoencapsulated carotenoids. *Food Bioprod. Process.* **2017**, *105*, 86–94. [CrossRef]
58. Limbo, S.; Torri, L.; Piergiovanni, L. Light-induced changes in an aqueous β-carotene system stored under halogen and fluorescent lamps, affected by two oxygen partial pressures. *J. Agric. Food Chem.* **2007**, *55*, 5238–5245. [CrossRef]
59. Kim, M.-J.; Bang, W.S.; Yuk, H.-G. 405 ± 5 nm light emitting diode illumination causes photodynamic inactivation of Salmonella spp. on fresh-cut papaya without deterioration. *Food Microbiol.* **2017**, *62*, 124–132. [CrossRef] [PubMed]
60. Pesek, C.; Warthesen, J. Photodegradation of carotenoids in a vegetable juice system. *J. Food Sci.* **1987**, *52*, 744–746. [CrossRef]

Article

Ultrasound-Assisted Extraction of Flavonoids from Kiwi Peel: Process Optimization and Bioactivity Assessment

Miguel Giordano [1,2], José Pinela [1], Maria Inês Dias [1], Ricardo C. Calhelha [1], Dejan Stojković [3], Marina Soković [3], Débora Tavares [4], Analía Laura Cánepa [2], Isabel C. F. R. Ferreira [1], Cristina Caleja [1,*] and Lillian Barros [1,*]

[1] Centro de Investigação de Montanha (CIMO), Instituto Politécnico de Bragança, Campus de Santa Apolónia, 5300-253 Bragança, Portugal; miguelgiordano21@hotmail.com (M.G.); jpinela@ipb.pt (J.P.); maria.ines@ipb.pt (M.I.D.); calhelha@ipb.pt (R.C.C.); iferreira@ipb.pt (I.C.F.R.F.)

[2] Centro de Investigación y Tecnología Química (CITeQ), UTN-CONICET, Facultad Regional Córdoba, Maestro López Esq, Cruz Roja Argentina s/n, Córdoba X5016ZAA, Argentina; acanepa@frc.utn.edu.ar

[3] Institute for Biological Research "Siniša Stanković"—National Institute of Republic of Serbia, University of Belgrade, Bulevar Despota Stefana 142, 11000 Belgrade, Serbia; dejanbio@yahoo.com (D.S.); marina.sokovic@mpn.gov.rs (M.S.)

[4] KiwiCoop, Rua Kiwicoop, n. 37, Vila Verde, 3770-305 Oliveira do Bairro, Portugal; dep.tecnico2@kiwicoop.com

* Correspondence: ccaleja@ipb.pt (C.C.); lillian@ipb.pt (L.B.)

Citation: Giordano, M.; Pinela, J.; Dias, M.I.; Calhelha, R.C.; Stojković, D.; Soković, M.; Tavares, D.; Cánepa, A.L.; Ferreira, I.C.F.R.; Caleja, C.; et al. Ultrasound-Assisted Extraction of Flavonoids from Kiwi Peel: Process Optimization and Bioactivity Assessment. *Appl. Sci.* **2021**, *11*, 6416. https://doi.org/10.3390/app11146416

Academic Editors: Gwiazdowska Daniela, Krzysztof Juś and Katarzyna Marchwińska

Received: 1 June 2021
Accepted: 8 July 2021
Published: 12 July 2021

Abstract: The nutritional quality of kiwifruit has been highlighted by several studies, while its peel is typically discarded as a by-product with no commercial value. This study was carried out to optimize the ultrasound-assisted extraction (UAE) of phenolic compounds from kiwi peel. Three independent variables (time (t), ultrasonic power (P) and ethanol concentration (EtOH)) were combined in a five-level central composite rotatable design coupled with the response surface methodology (RSM). The extraction yield determined gravimetrically and the content of phenolic compounds identified by HPLC-DAD-ESI/MS^n (namely two quercetin glycosides, one catechin isomer and one B-type (epi)catechin dimer) were the experimental responses used in the optimization. The polynomial models were successfully fitted to the experimental data and used to determine the optimal UAE conditions. The sonication of the sample at 94.4 W for 14.8 min, using 68.4% ethanol, resulted in a maximum of 1.51 ± 0.04 mg of flavonoids per g of extract, a result that allowed the experimental validation of the predictive model. The kiwi peel extract obtained under optimized conditions showed somehow promising bioactive properties, including antioxidant and antimicrobial effects, and no toxicity to Vero cells. Overall, this study contributes to the valorization of kiwi peel as a low-cost raw material for the development of natural ingredients (such as food preservatives) and also to the resource-use efficiency and circular bioeconomy.

Keywords: *Actinidia deliciosa*; by-product valorization; extraction optimization; bioactive properties; natural ingredients

1. Introduction

In the food industry, approximately 30% of the food produced each year is discarded worldwide. This represents 1300 million tons of food, one billion dollars in economic costs, 700 billion dollars in environmental costs and approximately 900 billion dollars in social costs [1]. Moreover, the peels, leaves, roots, tubers and seeds of fruits and vegetables that are discarded annually generate around 25% to 30% of the food industry waste [2]. These biowaste and by-products can cause environmental problems, such as aquatic life toxicity, surface and ground water contamination, changes in soil quality, greenhouse gases emissions, in addition to attracting disease vectors such as insects and rodents, among others [3]. Therefore, several alternatives have been studied to avoid this environmental scenario and obtain economic benefits through the valorization and recycling of the generated by-products [2].

The attempt to valorize plant by-products has aroused great interest in their recycling as a low-cost material for the development of bio-based ingredients with a high content of phenolic compounds, dietary fibers, minerals, and other phytochemicals [2,4]. These high added-value compounds have been correlated with health-promoting effects due to their antioxidant, anti-inflammatory, anti-mutagenic, and anti-diabetic properties, among others [5]. Today there is a large amount and diversity of agri-food by-products with an extremely variable composition, but its potential as sources of bioactive molecules is still little explored, being a niche of opportunities for obtaining high added-value molecules that can be introduced in the food cycle as natural food ingredients [5,6]. This approach contributes to the resource-use efficacy and circularity, as well as to the growing consumer' demand for food products formulated with alternative natural additives.

According to the literature, kiwi was originally a wild fruit native to China, but its commercial exploitation began in New Zealand with the species *Actinidia deliciosa*, of which there are around 75 varieties, "Hayward" being the best known [7]. Regarding nutritional value, this fruit is defined as a good source of nutrients, mainly fibers and minerals, as also of bioactive compounds such as polyphenols, vitamin C, and carotenoids, which display antioxidant properties [8]. Due to these compositional features, this fruit has been extensively studied, while the non-edible parts such as stems, leaves, peels, and seeds have been poorly investigated despite being a potential source of valuable compounds [9]. Thus, the sustainable use of these underused plant parts, especially its peel that is discarded as an industrial by-product with no commercial value, can contribute to its valorization and be achieved through its recycling into bioactive natural ingredients for exploitation by the agri-food sector [8,9].

Today, there are different non-conventional method that can be used to recover bioactive compounds from plant materials, such as the ultrasound-assisted extraction (UAE). During UAE, the physical forces developed by the acoustic cavitation caused by ultrasound waves promote the rupture of the plant tissues and the release of extractable compounds into the solvent in much less time than the conventional methods [2,4,10]. UAE is pointed out as a time-saving non-thermal method, presenting advantages over conventional maceration techniques involving temperature as intensification factor [11]. Therefore, this study was carried out to optimize the UAE of phenolic compounds from kiwi peel using the response surface methodology (RSM), and evaluate the in vitro bioactive properties of the extract obtained under optimized conditions in order to validate its potential to be used as a natural ingredient in the food industry.

2. Materials and Methods

2.1. Samples Preparation

Kiwi (*Actinidia deliciosa* cv. "Hayward") fruit samples were provided by KiwiCoop®, an organization of kiwi producers based in Oliveira do Bairro, Portugal. The fruit peel was separated from the pulp, lyophilized (FreeZone 4.5, Labconco, Kansas City, MO, USA) and stored under vacuum at −20 °C until further analysis.

2.2. Experimental Design for Extraction Optimization

A central composite rotatable design (CCRD) combining five-levels of the independent variables X_1 (time, t, 1–45 min), X_2 (ultrasonic power, P, 5–500 W) and X_3 (ethanol concentration, EtOH, 0–100%, *v/v*) was implemented to optimize the extraction of flavonoids from kiwi peel using RSM (Supplementary Material Table S1). These variables and the respective range of values were selected based on previous optimization studies [11,12]. Design-Expert software, Version 11 (Stat-Ease, Inc., Minneapolis, MN, USA) was used to generate the 20 experimental points of the CCRD design, which included eight factorial points, six axial or star points chosen to allow rotatability, and six replicated center points. The 20 runs were randomized to minimize the effects of unexpected variability.

2.3. Ultrasound-Assisted Extractions (UAE)

The UAE was performed using an ultrasonic system (QSonica sonicators, model CL-334, Newtown, CT, USA) equipped with a titanium probe. A known sample weight (1.5 g) was mixed with 50 mL of solvent (0–100%) and sonicated at 5–500 W (at 20 kHz frequency) for 1–45 min. The temperature (~25 °C) and solid/liquid ratio (30 g/L) were kept constant. After processing, the mixtures were centrifuged at 450 rpm for 5 min and filtered through Whatman no. 4 filter paper. An aliquot of each filtrate was filtered through syringe filter discs for high-performance liquid chromatography (HPLC) analysis of flavonoids, another was used for determination of the extraction yield (extract weight or extracted solids, %, *w/w*) by gravimetry, and the remaining filtrate was lyophilized for further evaluation of bioactive properties.

2.4. Identification and Quantification of Phenolic Compounds

For analysis of phenolic compounds, a UPLC Dionex Ultimate 3000 system was used (Thermo Scientific, San Jose, CA, USA) following the analytical procedure previously described by Bessada et al. [13]. Detection was performed with a diode array detector (DAD) and a Linear Ion Trap (LTQ XL) mass spectrometer (Thermo Finnigan, San Jose, CA, USA) working in negative mode and equipped with an electrospray ionization (ESI) source. Chromatographic separation was made on a Waters Spherisorb S3 ODS-2 column (3 μm, 4.6 mm × 150 mm; Waters, Milford, MA, USA). The identification was achieved as previously described [13]. For quantitative analysis, 7-level calibration curves were constructed based on the UV-Vis signal of the quercetin-3-*O*-glucoside ($y = 21719x + 88805$; $r^2 = 0.9994$) and (-)-catechin ($y = 8387.1x + 71124$; $r^2 = 0.9964$) standards. The analysis was performed in triplicate and the results were expressed in mg per g of dry weight (dw).

2.5. Extraction Process Modelling and Statistical Analysis

The dependent variables Y_1 (extraction yield, %, *w/w*), Y_2 (B-type (epi)catechin dimer content, mg/g), Y_3 (epicatechin content, mg/g), Y_4 (quercetin-3-*O*-glucoside content, mg/g), Y_5 (quercetin-3-*O*-rhamnoside content, mg/g), and Y_6 (total flavonoids content, mg/g) were used in the extraction process optimization. The response surface models were fitted by means of least squares calculation using the following second-order polynomial equation:

$$Y = b_0 + \sum_{i=1}^{n} b_i X_i + \sum_{\substack{i=1 \\ j>i}}^{n-1} \sum_{j=2}^{n} b_{ij} X_i X_j + \sum_{i=1}^{n} b_{ii} X_i^2 \quad (1)$$

where Y corresponds to the dependent variable to be modelled, X_i and X_j define the independent variables, b_0 is the constant coefficient, b_i is the coefficient of the linear effect, b_{ij} is the coefficient of the interaction effect, b_{ii} is the coefficient of the quadratic effect, and n is the number of variables ($n = 3$).

Fitting procedures, coefficient estimates, and statistical analysis were performed using the Design-Expert software. Analysis of variance (ANOVA) was used to assess the significance of the models and of all the terms that make up the models, as well as the lack-of-fit. Only the statistically significant terms ($p < 0.05$) were used in the models' construction (except those required to ensure hierarchy). Coefficient of determination (R^2), adjusted coefficient of determination (R^2_{adj}), and adequate precision were used to estimate the adequacy of the polynomial equation to the response. The lack-of-fit measures the quality of the model's fit to the experimental data; thus, it must be non-significant ($p > 0.05$).

2.6. Models Validation and Evaluation of the Bioactivity of the Extract Produced under Optimized Extraction Conditions

The optimized global UAE conditions that maximize the recovery of flavonoids from kiwi peel were applied to obtain a flavonoid-rich extract, following the procedure described above. This extract was used for experimental validation of the theoretical models, performed through the analysis of the experimental responses (extraction yield and flavonoid content) of this new extract and comparison with the model-predicted values. The in vitro bioactivity of this extract was also evaluated as described below.

2.7. Bioactivities Evaluation

2.7.1. Antioxidant Activity

To evaluate the extract capacity to inhibit the formation of thiobarbituric acid reactive substances (TBARS), an in vitro assay based on the monitoring of malondialdehyde (MDA)-TBA complexes was implemented as previously reported [14]. Porcine brain cells were used as biological substrates. The results were expressed as IC_{50} values (mg/mL).

The extract capacity to protect sheep erythrocytes from oxidative haemolysis was tested by the OxHLIA assay [14]. Briefly, an erythrocyte solution (2.8%, *v/v*) prepared in phosphate-buffered saline (PBS, pH 7.4) was mixed with either: (i) extract solution in PBS, (ii) Trolox in PBS, (iii) PBS (control), or (iv) water (100% haemolysis). After pre-incubation at 37 °C for 10 min with shaking, 2,2′-azobis (2-amidinopropane) dihydrochloride (AAPH) (160 mM in PBS) was added, and the optical density was measured at 690 nm over time until complete haemolysis. The results were given as IC_{50} values (mg/mL) at a 60 min Δt.

2.7.2. Anti-Inflammatory Activity

The anti-inflammatory activity was evaluated by the extract capacity to inhibit the nitric oxide (NO) production by lipopolysaccharide (LPS)-stimulated RAW 264.7 cells, following a protocol previously described by Barros et al. [15]. Dexamethasone was used as positive control. The results were expressed as IC_{50} values (μg/mL).

2.7.3. Cytotoxicity to Tumor Cell Lines and Hepatotoxic Activity

The extract capacity to inhibit the cell growth was screened against human tumor cell lines, namely MCF-7 (breast adenocarcinoma), NCI-H460 (non-small cell lung cancer), AGS (gastric adenocarcinoma), and CaCo-2 (colorectal adenocarcinoma), and the normal African Green Monkey kidney epithelial Vero cell line. The Sulforhodamine B assay was followed as previously described [16]. Ellipticine was used as positive control. The results were expressed as GI_{50} values (μg/mL).

2.7.4. Antimicrobial Activity

The antibacterial activity was evaluated following the methodology previously described by Soković et al. [17] against the Gram-negative bacteria *Escherichia coli* (ATCC 25922), *Enterobacter cloacae* (ATCC 35030), and *Salmonella* Typhimurium (ATCC 13311), and the Gram-positive bacteria *Listeria monocytogenes* (NCTC 7973), *Staphylococcus aureus* (ATCC 11632), and *Bacillus cereus* (food isolate). The results were given as minimum inhibitory and bactericidal concentrations (MIC and MBC, respectively).

For antifungal activity, the methodology described by Soković and van Griensven [18] was implemented and the microfungi *Aspergillus ochraceus* (ATCC 12066), *Aspergillus niger* (ATCC 6275), *Aspergillus versicolor* (ATCC 11730), *Penicillium funiculosum* (ATCC 36839), *Penicillium aurantiogriseum* (food isolate), and *Trichoderma viride* (IAM 5061) were tested. MIC and minimum fungicidal concentration (MFC) were determined.

All microorganisms were obtained from the Mycological Laboratory, Institute for Biological Research "Sinisa Stanković", University of Belgrade (Belgrade, Serbia). Sodium benzoate (E211) and potassium metabisulfite (E224) were used as positive controls.

3. Results and Discussion

The extraction of polyphenols from plant materials is affected by different factors related to the compositional and structural nature of the plant material (which must be reduced to a small particle size to increase the sample-to-solvent contact area) and to the factors applied during the extraction process, such as solvent type, temperature, ultrasonic power, solid/liquid ratio, and processing time. The selection of extraction methods and the determination of processing conditions that maximize the recovery of bioactive compounds has gained particular interest in recent years due to the current trend to valorize and recycle agri-food by-products [2,4]. In this sense, efforts have been made to develop more efficient and sustainable extraction processes, capable of improving extraction yield and selectivity [11,19,20]. However, the extrapolation of results obtained with different natural matrices or using different extraction techniques can be a difficult or wrong task. Therefore, in this study, the suitability of UAE to recover flavonoids from kiwi peel was investigated and optimized by RSM.

3.1. Phenolic Profile of the Kiwi Peel Extract and Experimental Data for UAE Optimization

Kiwi is rich in polyphenols with antioxidant activity and often described as a "super-fruit" due to its low caloric value and high amount of water, dietary fiber, and vitamin C, among other nutrients [21,22]. The HPLC-DAD-ESI/MS^n (Table 1) analysis allowed the tentative identification of two quercetin glycosides, namely quercetin-3-*O*-glucoside (identified based on the pseudomolecular ion $[M-H]^-$ at *m/z* 463 and the fragment ion at *m/z* 301 and by comparison with the available standard compound) and quercetin-3-*O*-ramnoside (pseudomolecular ion $[M-H]^-$ at *m/z* at 477 and fragment ion at *m/z* 301), as also two flavan-3-ols, namely epicatechin (identified based on the pseudomolecular ion $[M-H]^-$ at *m/z* 289 and fragments at *m/z* 245, 205, and 179), and a B-type (epi)catechin dimer (pseudomolecular ion $[M-H]^-$ at *m/z* 577 and fragments at *m/z* 451, 425, 407, and 289). A representative HPLC phenolic profile of the kiwi peel extracts is present in Supplementary Materials Figure S1.

Table 1. Flavonoids composition of the kiwi peel extract. The retention time (Rt), wavelengths of maximum absorption in the visible region (λ_{max}), and mass spectral data are presented.

Peak	Rt (min)	λ_{max} (nm)	$[M-H]^-$ (*m/z*)	MS^2 (*m/z*)	Tentative Identification
1 A	5.85	281	577	451(27), 425(100), 407(32), 289(10)	B-type (epi)catechin dimer
2 A	7.63	280	289	245(100), 205(41), 179(17)	Epicatechin
3 B	18.71	352	463	301(100)	Quercetin-3-*O*-glucoside
4 B	20.28	351	477	301(100)	Quercetin-3-*O*-rhamnoside

Calibration curves used in the quantification: (A) catechin ($y = 8387.1x + 71{,}124$, $r^2 = 0.9964$) and (B) quercetin-3-*O*-glucoside ($y = 21719x + 88{,}805$, $r^2 = 0.9994$).

The results of the 20 experimental runs of the CCRD design applied to optimize the UAE of flavonoids from the kiwi peel are presented in Table 2. The extraction yield ranged from approximately 37% to 59% with runs 14, involving 100% ethanol and medium levels of the other two independent variables, and 12 and 4, which combined 500 W power with medium levels (0) of the other variables and medium-high levels of time and ultrasonic power with a medium-low ethanol concentration, respectively. In turn, the total content of flavonoids ranged from 1.00 to 1.75 mg/g of extract and the lowest values were obtained with runs 9 and 10, corresponding to the use of the lowest and highest levels (α values) of the variable time, mainly due to the low levels of epicatechin achieved with these runs. In general, the recovery of the B-type (epi)catechin dimer appeared to have been promoted by the two axial points corresponding to the extraction solvent (runs 13 and 14), while both quercetin glycosides were better extracted with the run 10.

Table 2. Experimental responses obtained under the extraction conditions defined by the CCRD design matrix for the extraction yield and flavonoids content.

Runs	Experimental Domain			Experimental Responses *					
	t (min)	P (W)	EtOH (%)	Y_1 (%, w/w)	Y_2 (mg/g dw)	Y_3 (mg/g dw)	Y_4 (mg/g dw)	Y_5 (mg/g dw)	Y_6 (mg/g dw)
1	10 (−1)	106 (−1)	20 (−1)	47.90	0.5389	0.6237	0.1411	0.1380	1.4416
2	36 (+1)	106 (−1)	20 (−1)	50.63	0.4629	0.8580	0.1440	0.1448	1.6097
3	10 (−1)	400 (+1)	20 (−1)	54.09	0.4954	0.4684	0.0601	0.1385	1.1624
4	36 (+1)	400 (+1)	20 (−1)	59.14	0.4873	0.7263	0.0601	0.1388	1.4126
5	10 (−1)	106 (−1)	80 (+1)	45.09	0.5013	0.4366	0.1415	0.1421	1.2214
6	36 (+1)	106 (−1)	80 (+1)	45.66	0.4892	0.4393	0.1430	0.1423	1.2138
7	10 (−1)	400 (+1)	80 (+1)	47.35	0.5249	0.4374	0.1442	0.1409	1.2475
8	36 (+1)	400 (+1)	80 (+1)	51.10	0.5126	0.3645	0.1451	0.1429	1.1651
9	1 (−1.68)	253 (0)	50 (0)	50.44	0.4215	0.2826	0.1498	0.1462	1.0001
10	45 (+1.68)	253 (0)	50 (0)	51.96	0.3686	0.3884	0.1622	0.1493	1.0685
11	23 (0)	5 (−1.68)	50 (0)	50.43	0.4304	1.0289	0.1446	0.1464	1.7502
12	23 (0)	500 (+1.68)	50 (0)	59.02	0.4258	1.2575	0.0833	0.1487	1.6300
13	23 (0)	253 (0)	0 (−1.68)	48.35	0.5983	0.6216	0.0851	0.1329	1.4379
14	23 (0)	253 (0)	100 (+1.68)	36.87	0.6038	0.3010	0.1378	0.1361	1.1787
15	23 (0)	253 (0)	50 (0)	52.02	0.3067	1.1225	0.1565	0.1471	1.7328
16	23 (0)	253 (0)	50 (0)	46.13	0.2932	0.9786	0.1518	0.1449	1.5685
17	23 (0)	253 (0)	50 (0)	49.73	0.3074	1.0002	0.1535	0.1459	1.6070
18	23 (0)	253 (0)	50 (0)	50.58	0.3172	1.0508	0.1547	0.1468	1.6695
19	23 (0)	253 (0)	50 (0)	50.02	0.3068	1.0231	0.1536	0.1455	1.6290
20	23 (0)	253 (0)	50 (0)	51.38	0.3189	1.0933	0.1592	0.1474	1.7189

* It is presented the mean value of three determinations. Y_1: extraction yield; Y_2: B-type (epi)catechin dimer; Y_3: epicatechin; Y_4: quercetin-3-*O*-glucoside; Y_5: quercetin-3-*O*-rhamnoside; Y_6: total content of flavonoids.

3.2. Models Fitting and Statistical Verification

The response values in Table 2 were fitted to the second-order polynomial Equation (1) using the Design-Expert software, but just the significant parameters (assessed at a 95% confidence level) were used in the development of the theoretical models. The results of ANOVA and regression analyses are presented in Supplementary Materials Table S2. The developed polynomial models, expressed in coded values, are presented in Equations (2)–(7):

$$Y_1 = 50.4 + 1.1t + 2.7P - 3.1S + 1.8P^2 - 2.5S^2 \quad (2)$$

$$Y_2 = 0.308 - 0.014t + 0.002P + 0.002S + 0.034t^2 + 0.046P^2 + 0.107S^2 \quad (3)$$

$$Y_3 = 1.04 + 0.04t - 0.11S - 0.07tS - 0.26t^2 - 0.22S^2 \quad (4)$$

$$Y_4 = 0.155 - 0.019P + 0.002S + 0.021PS - 0.015P^2 - 0.016S^2 \quad (5)$$

$$Y_5 = 0.1464 + 0.0011t + 0.001S - 0.0046S^2 \quad (6)$$

$$Y_6 = 1.66 + 0.03t - 0.05P - 0.09S - 0.06tS + 0.06PS - 0.22t^2 - 0.13S^2 \quad (7)$$

In the mathematical models developed for each response variable, the coefficients of the terms t, P, and S illustrate the effect of the independent variables time, ultrasonic power, and ethanol concentration, respectively, and their interactions. Since the expected effects on the response are denoted by the parametric values, the higher the parametric value, the more significant the weight of the respective variable will be, regardless of its sign. For interactions, synergistic effects are denoted by a positive sign, while an antagonistic interaction between variables is translated by a negative sign [23]. In each model Equations (2)–(7), the intercept corresponds to the overall average response of the 20 experimental runs of the CCRD design in Table 2. Based on the intercept values, epicatechin stands out as the main phenolic compound (corresponding to almost 63% of the total content) identified in the kiwi peel by-product, followed by B-type (epi)catechin dimer (18.6%), and the two quercetin glycosides (18.2%) (Supplementary Materials Table S2).

All Equations (2)–(7) presented a non-significant lack-of-fit ($p > 0.05$) and an adequate precision greater than 22.4, which indicates that the theoretical models adequately describe the effects of the independent variables on the target responses [24]. The coefficients R^2 and R^2_{adj} were higher than 0.888 and 0.864 in all cases, respectively (Supplementary Material Table S2), indicating that the variability of each response can be explained by the independent variables involved in the extraction. In addition, values $\geq$ 22.49 were obtained for adequate precision, which is a measure of signal-to-noise ratio that compares the range of the predicted values at the design points to the average prediction error. High accuracy was also demonstrated by the low values of the coefficient of variance. Thus, the developed theoretical models were statistically validated and used in the following steps to predict the optimal UAE conditions for recovery of polyphenols from kiwi peel.

Certain peculiarities regarding the overall effects of the independent variables on the UAE of flavonoids from kiwi peel can be inferred from the complexity of the model equations. Based on Equation (2) and Supplementary Material Table S2, it can be settled that the extraction yield was significantly affected by the three independent variables involved in the extraction. The ethanol concentration was the most relevant process variable, effecting the extraction through negative linear and quadratic effects. The ultrasonic power ranked second and caused positive linear and quadratic effects, while the extraction time only induced a linear effect. No integrations between variables were observed for the extraction yield. In turn, Equation (7) translates the complexity of the extraction trend over the total flavonoid content, where it is interesting to highlight the negative quadratic effects of the variables time and ethanol concentration, as well as the negative and positive interactions of ethanol concentration with time and ultrasonic power, respectively. This result supports the use of RSM as optimization tool, since the one-factor-at-a-time approaches do not assess interactive effects. The individual flavonoid compounds appeared to have been somewhat differently affected, mainly B-type (epi)catechin dimer.

3.3. Effect of the Extraction Parameters on the Responses

The 3D response surface graphs constructed to illustrate the effect of the independent variables involved in the UAE of flavonoids from kiwi peel are presented in Figure 1. In each graph, the excluded variable was positioned at its individual optimal value shown in Table 3. The B-type (epi)catechin dimer showed an opposite extraction tendency compared to that of the other flavonoids.

As observed in Figure 1, the increase in the value of the three variables up to intermediate levels decreased the extraction of this proanthocyanidin, but the consequent increase induced a new improvement in its recovery rate. The extraction trend thus followed quadratic effects, which were more pronounced for EtOH > P > t. In this particular case, the central projections of the surfaces at the base of each 3D graph illustrate the variable ranges where the extraction of B-type (epi)catechin dimer was lower. Regarding epicatechin, its extraction was promoted by the increase in processing time and ethanol concentration up to around 25 min and 42% EtOH, respectively, but the consequent increased caused a reduction in the recovery yield, probably due to its degradation. Actually, the red colored response surface area of the 3D graph of the combined effects of these two variables illustrates very well the optimal response (Figure 1). Contrariwise, the ultrasonic power did not significantly affect ($p > 0.05$) the extraction of this antioxidant flavonoid (Supplementary Material Table S2). Based on these observations and the reverse response surfaces of B-type (epi)catechin dimer and epicatechin (Figure 1), it may be suggested that the UAE process may have promoted the breakdown of this B-type proanthocyanidin into its flavanol subunits.

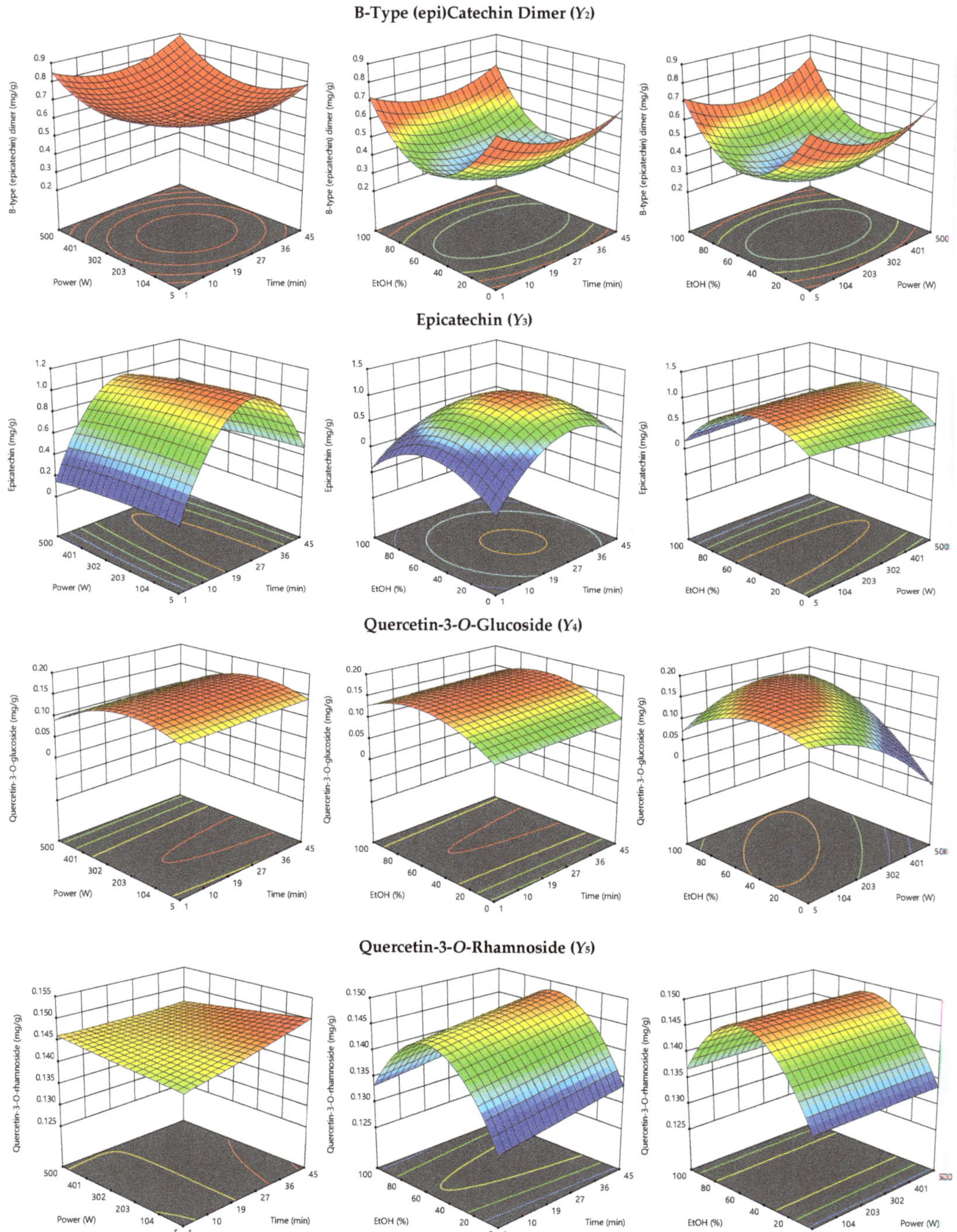

Figure 1. Response surface graphs for the combined effects of the independent variables on the content (mg/g) of individual flavonoids (Y_2–Y_5) recovered from kiwi peel. In each graph, the excluded variable was fixed at its optimal value.

Table 3. Optimal processing conditions that maximize the extraction of flavonoids from kiwi peel and model-predicted and experimental response optimum.

	Optimal Processing Conditions			Response Optimum	
	Time (min)	Power (W)	EtOH (%)	Model-Predicted Values	Experimental Values
Individual conditions for each response variable					
Extraction yield (extract)	34.4	483.0	34.1	61 ± 1% (*w/w*)	-
B-type (epi)catechin dimer	11.2	393.0	94.8	0.64 ± 0.01 mg/g dw	-
Epicatechin	24.6	222.6	41.6	1.06 ± 0.02 mg/g dw	-
Quercetin-3-*O*-glucoside	39.2	191.4	59.2	0.164 ± 0.002 mg/g dw	-
Quercetin-3-*O*-rhamnoside	45.0	257.6	53.2	0.148 ± 0.001 mg/g dw	-
Total flavonoids	24.9	5.0	34.1	1.82 ± 0.03 mg/g dw	-
Global conditions considering all response variables					
Extraction yield (extract)	14.8	94.4	68.4	46 ± 1% (*w/w*)	46 ± 2% (*w/w*)
B-type (epi)catechin dimer				0.426 ± 0.008 mg/g dw	0.432 ± 0.006 mg/g dw
Epicatechin				0.78 ± 0.02 mg/g dw	0.78 ± 0.04 mg/g dw
Quercetin-3-*O*-glucoside				0.148 ± 0.002 mg/g dw	0.150 ± 0.003 mg/g dw
Quercetin-3-*O*-rhamnoside				0.145 ± 0.001 mg/g dw	0.1468 ± 0.0002 mg/g dw
Total flavonoids				1.49 ± 0.03 mg/g dw	1.51 ± 0.04 mg/g dw

For the two quercetin glycosides, the 20 experimental runs of the CCRD design yielded comparable mean responses, as shown by the intercession values in Equations (5)–(6) and Supplementary Material Table S2. The longer the processing time, the greater the linear recovery of these quercetin glycosides, but this variable was significant ($p < 0.05$) only for quercetin-3-*O*-rhamnoside. In turn, the ultrasonic power merely impacted the extraction of quercetin-3-*O*-glucoside, whose yield increased with the application of up to ~190 W and then decreased for higher powers. Both compounds were significantly affected by the extraction solvent ($p < 0.05$); while the recovery of quercetin-3-*O*-glucoside mainly followed a linear effect, quadratic effects marked the trend of quercetin-3-*O*-rhamnoside extraction. Ethanol/water ranging from 50 to 60% appeared to be preferable mixtures. In the particular case of quercetin-3-*O*-glucoside, it is also interesting to note the strong positive interaction between ultrasonic power and ethanol concentration.

Figure 2 illustrates the response surface graphs for the effects of the independent variables on the total content of flavonoids and extraction yield obtained from kiwi peel. The excluded variable in each 3D graph was fixed at its optimal value. In general, the extraction of the total flavonoids content follows a trend comparable to that discussed above for epicatechin, with additional interactive effects between the solvent and the other two variables. So, the lower the ethanol concentration and ultrasonic power (EtOH × P), the greater the recovery of flavonoids, while lower ethanol concentrations combined with longer processing times (EtOH × t) seem to improve the UAE process efficiency.

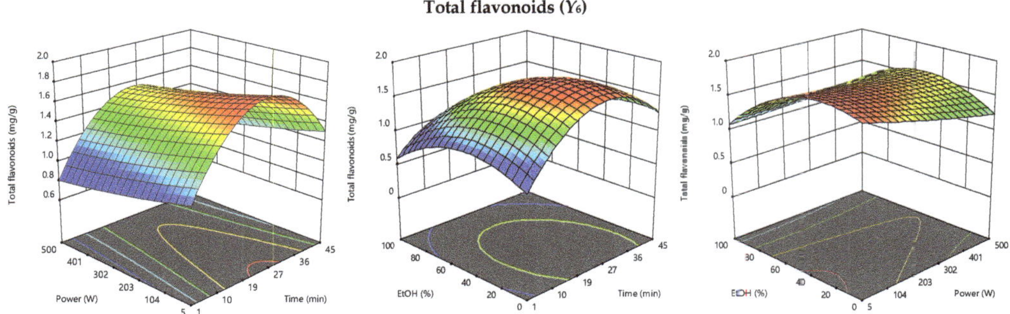

Figure 2. *Cont.*

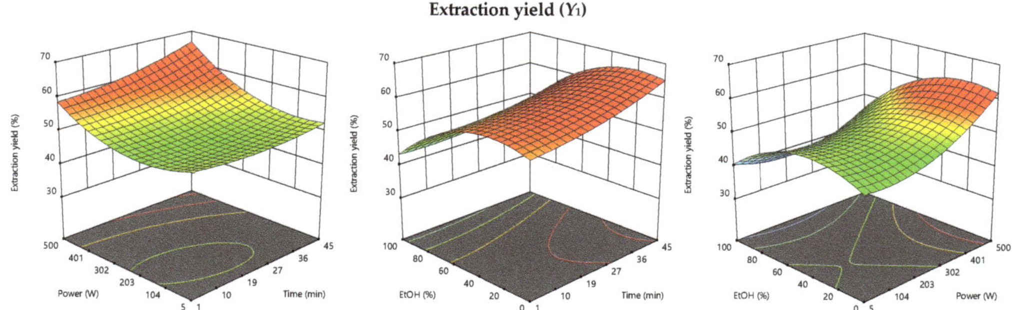

Figure 2. Response surface graphs for the combined effects of the independent variables on the total content of flavonoids (Y_6) and extraction yield (Y_1) obtained from kiwi peel. In each graph, the excluded variable was fixed at its optimal value.

The extraction solvent was the variables that most affected the extraction yield and the highest extract weights were obtained when sonicating the kiwi peel sample with lower ethanol concentrations (Figure 2). A higher ultrasonic power and longer extraction times also favored the recovery rate. However, it should be noted that the increase in the extract weight does not necessarily resulted in a greater extraction of flavonoids. As seen in Figures 2 and 3, although the more intense conditions of time and ultrasonic power have led to a greater extract weight, these have negatively affected the total flavonoids content. Therefore, compounds other than flavonoids were being extracted (possibly fibers and other carbohydrates, given the intrinsic nature of the kiwi peel) and represented the major fraction of the obtained extract. The optimal points of these two responses are also shown in the cube plots in Supplementary Materials Figure S2A,B, respectively, which shows how the three variables combine to affect the response. A flavonoid-rich extract can thus be obtained more selectively through the control of these two process variables (t and P). Interestingly, the ethanol percentage that led to a greater amount of from kiwi peel extract also favored the recovery of flavonoids (Figures 2 and 3 and Supplementary Materials Figure S2).

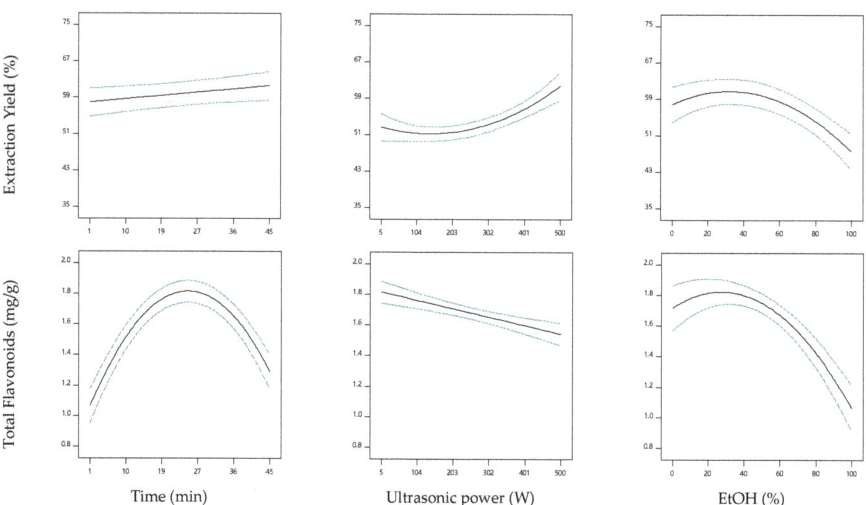

Figure 3. 2D response graphs for the effects of the independent variables on the total content of flavonoids (Y_6) and extraction yield (Y_1) obtained from kiwi peel. In each graph, the excluded variables were fixed at their optimal value.

3.4. Optimal UAE Conditions

For numerical optimization of the UAE process, the Design-Expert software used searches for a combination of factor levels that simultaneously met the requirements placed on each dependent and independent variable. The optimization required that goals be defined for both types of variables, in order to combine all goals into a desirability function. To find a good set of conditions that met the desired goals, the three independent variables were set within the experimental range, while the response or dependent variable was set at maximum. In addition, equal "importance" of goals was given to the variables. The model-predicted UAE conditions that maximize each individual response to optimal values are presented in Table 3. The maximum extraction yield of 61% was obtained by processing the kiwi peel sample at 483 W for 34.4 min with 34% ethanol. In turn, 1.82 mg/g dw of total flavonoids can be reached by sonicating the powdered sample at a power of only 5 W for 25 min, using 69% ethanol. While B-type (epi)catechin dimer was better extracted with a mere 11 min processing with a high power of 393 W and 95% ethanol, epicatechin required 24 min sonication at 223 W, using 42% ethanol. The quercetin glycosides required a longer sonication time of 39–45 min at 191–258 W using 53–59% ethanol, yielding 0.312 mg/g dw of both compounds.

Since for the industrial sector interested in bioactive plant extracts it is important to obtain a high amount of both extract and bioactive compounds through sustainable processes, global UAE conditions that simultaneously maximize all responses were also determined (Table 3 and Supplementary Materials Figure S2C). Based on this second optimization step, 15 min sonication at 94.4 W, using 68.4% ethanol as solvent, were found to be the optimal conditions to simultaneously maximize the response variables as much as possible (46 ± 1% extraction yield and 1.49 ± 0.03 mg/g dw of flavonoids).

3.5. Experimental Validation of the Predictive Model

The global UAE conditions that maximize both the extraction yield and the recovery of flavonoids from kiwi peel (Table 3) were experimentally tested in triplicate to evaluate the predictive accuracy of the model and to obtain a flavonoid-rich extract for evaluation of in vitro bioactive properties. As shown in Table 3, the experimental data were in good agreement with the model-predicted values, as confirmed by the post-analysis verification performed in Design-Expert software (α = 0.05). The UAE yielded 46 ± 2% of crude extract, a value that did not differ from the predicted 46 ± 1%, and 1.51 mg/g dw of total flavonoids. The predictive capacity of the model was thus experimentally validated.

3.6. Bioactivity of the Kiwi Peel Extract Obtained under Optimized UAE Conditions

3.6.1. Antioxidant Activity

The kiwi peel extract obtained under the global UAE conditions (Table 3) was tested for its capacity to inhibit the formation of thiobarbituric acid reactive substances (TBARS) and the oxidative hemolysis (OxHLIA). The results were expressed in IC_{50} values, which correspond to the extract concentration that provides 50% of antioxidant activity in the TBARS assay or required to protect 50% of the erythrocyte population from the free radicals generated by AAPH for a Δt of 60 min. Therefore, the lower the IC_{50} values, the greater the antioxidant activity [14]. The extract showed IC_{50} values of 0.37 ± 0.02 mg/mL and 1.01 ± 0.01 mg/mL for TBARS and OxHLIA, respectively, while the trolox presented a considerably higher activity (IC_{50} values of 5.4 ± 0.3 μg/mL and 21.8 ± 0.3 μg/mL, respectively). However, trolox is a pure antioxidant compound whereas the extract may contain other constituents without bioactivity, which justify the observed differences.

Kiwifruit has been described as having a high antioxidant capacity [3], but some studies on its peel have also been carried out. Dias et al. [25] compared the bioactive potential of the pulp and peel of two kiwi varieties and highlight the green kiwi peel as the sample with greater antioxidant activity than the others, including the pulps. These results are in agreement with those of Bernardes et al. [26] and Fiorentino et al. [27], who described higher values of antioxidant activity for kiwi peel than for pulp. The same conclusion was reached by Soquetta [28], who compared flours obtained from by-products (epicarp and bagasse) of two kiwi varieties, which observed a greater antioxidant activity in the flours of the peel than in those of bagasse. The difference in the IC_{50} values described in these studies can be justified by the different kiwi varieties or by the different extraction methodologies and solvents applied, as well as by the treatment performed to obtain the plant sample. Considering that the extract evaluated in the present work was obtained under optimized conditions that aimed at maximizing the flavonoids content, a better performance was expected in the in vitro assays.

3.6.2. Cytotoxic and Anti-Inflammatory Activity

The cytotoxic activity of the kiwi peel extract against tumor (MCF-7, NCI-H460, AGS and CaCo-2) and non-tumor (Vero) cell lines was evaluated, as well as its anti-inflammatory activity via NO production inhibition. The extract showed activity against NCI-H460 (non-small cell lung cancer) cells, with an GI_{50} of 309 ± 16 μg/mL. However, no activity was observed against other cell lines (MCF-7, AGS and CaCo-2) at the tested concentrations (GI_{50} > 400 μg/mL). Despite these results, Moita [29] attributed cytotoxic effects to kiwi peel and pulp extracts against HepG2 (hepatocellular carcinoma) and CaCo-2 cells. Dias et al. [25] also demonstrated that kiwi peel hydroethanolic extracts have cytotoxicity to MCF-7, NCI-H460, HepG2 and HeLa (cervical carcinoma) cells. In turn, Lim et al. [30] investigated the anti-proliferative effects of kiwi extracts against HepG2, HT29 (colon carcinoma) and LoVo (colon carcinoma) tumor cells and observed that the extracts did not effectively inhibit cell proliferation. The absence of toxicity to non-tumor cells was herein demonstrated with the Vero cells (GI_{50} > 400 μg/mL). The kiwi peel extract also did not show anti-inflammatory activity, contrary to what has been described in the studies by An et al. [31] and Dias et al. [25].

The difference observed between the activities presented by our sample and the literature data may be explained by the fact that the data were obtained from different plant parts, by differences in the adopted analytical and extraction methodologies and in the used proportions of solvents, but also because they are from different varieties or species of the same genus. Therefore, further studies using different human tumor cell lines and purified compounds will be needed to confirm the bioactivity of the kiwi peel compounds.

3.6.3. Antimicrobial Activity

The antimicrobial activity of the kiwi peel extract was tested against some bacteria, namely *Bacillus cereus*, *Staphylococcus aureus*, *Listeria monocytogenes*, *Escherichia coli*, *Enterobacter cloacae* and *Salmonella* Typhimurium, and some fungi, namely *Aspergillus ochraceus*, *Aspergillus niger*, *Aspergillus versicolor*, *Penicillium funiculosum*, *Penicillium aurantiogriseum* and *Trichoderma viride*. To compare the effectiveness of the kiwi peel extract, two synthetic food preservatives, sodium benzoate (E211) and potassium metabisulfite (E224), were used as positive controls. Table 4 shows the obtained values of minimum inhibitory and bactericidal or fungicidal concentration (MIC, MBC and MFC, respectively). As shown in Table 4, *Staphylococcus aureus* and *Escherichia coli* were the most sensitive bacteria to the tested extract and *Aspergillus ochraceus*, *Aspergillus versicolor*, *Penicillium funiculosum* and *Trichoderma viride* were the most sensitive fungi.

For antimicrobial activity, the MIC and MBC values show that the kiwi peel extract behaves very similar to the synthetic preservative E224, when it was used against Gram-positive bacteria, such as *Bacillus cereus* and *Staphylococcus aureus*, but adopts a behavior similar to the preservative synthetic E211, when tested against Gram-negative bacteria, such as *Escherichia coli* and *Enterobacter cloacae*. Regarding the bacteria *Listeria monocytogenes* and *Salmonella* Typhimurium, the extract was less effective compared to the respective controls. In general, the kiwi peel extract appears to have a greater effect against fungi compared to bacteria, with lower MIC and MFC values. In addition, the extracts showed, for most of the microorganisms, better or equal activity than both synthetic preservatives.

Table 4. Antimicrobial activity of the kiwi peel extract obtained under optimized UAE conditions.

	Kiwi Peel Extract		E211		E224	
Antibacterial activity	MIC	MBC	MIC	MBC	MIC	MBC
Staphylococcus aureus	1	2	4	4	1	1
Bacillus cereus	2	4	0.5	0.5	2	4
Listeria monocytogenes	2	4	1	2	0.5	1
Escherichia coli	1	2	1	2	0.5	1
Salmonella Typhimurium	2	4	1	2	1	1
Enterobacter cloacae	2	4	2	4	0.5	0.5
Antifungal activity	MIC	MFC	MIC	MFC	MIC	MFC
Aspergillus ochraceus	0.5	1	1	2	1	1
Aspergillus niger	1	2	1	2	1	1
Aspergillus versicolor	0.5	1	2	2	1	1
Penicillium funiculosum	0.5	1	1	2	0.5	0.5
Penicillium aurantiogriseum	1	2	2	4	1	1
Trichoderma viride	0.5	0.5	1	2	0.5	0.5

E211: sodium benzoate; E224: potassium metabisulfite; MIC: minimum inhibitory concentration; MBC: minimum bactericidal concentration; MFC: minimum fungicidal concentration.

These results are in agreement with studies that describe the antimicrobial potential of *Actinide delicious* extracts against bacterial and fungal growth, which may be useful to help prevent food deterioration processes and the occurrence of possible contamination [25]. Kichaoi et al. [32] studied the antimicrobial potential of kiwi, pomegranate and grapefruit peels, and found that, although all fruit peels have great antimicrobial power, in most cases the kiwi peel stood out from the other samples, regardless of the type of extraction and solvent used. Soquetta [28] reported the antimicrobial activity of kiwi peel and bagasse flours against *Salmonella* sp., *Bacillus cereus* and *Staphylococcus aureus*. In a study on the isolation and characterization of an allergenic kiwi protein *in natura*, Gavrovic-Jankulovic et al. [33] verified its antifungal activity against *Saccharomyces carlsbergensis* and *Candida albicans*. Furthermore, Xia and Ng [34] verified the antifungal activity of fresh kiwi against *Fusarium oxysporum*. Therefore, the results presented in this study are somewhat in agreement with the results previously reported in the literature. It can also be concluded that the observed effects could be attributed not only to phenolic compounds, but also to other bioactive constituents of this plant by-product.

4. Conclusions

The chromatographic analysis allowed identifying four flavonoids in kiwi peel extracts, namely B-type (epi)catechin dimer, epicatechin, quercetin-3-O-glucoside, and quercetin-3-*O*-rhamnoside. In order to valorize this by-product as a source of bioactive compounds, a five-level CCRD design coupled to RSM was successfully implemented. The UAE was significantly affected by the independent variables time, ultrasonic power and ethanol concentration, and the theoretical models, fitted to the experimental data by means of least squares calculation using a second-order polynomial equation, were validated based on different statistical criteria. Under the optimized overall UAE condition, it was possible to obtain 46% extract weight and 1.51 mg/g dw of flavonoids. After experimental validation of the predictive model, the bioactivity of the kiwi peel extract obtained under the optimal UAE conditions was evaluated in vitro, showing a somehow promising bioactive potential. Thus, this study contributed to the valorization of kiwi peel through its recycling into a bioactive extract with potential for application in the food industry as a natural preservative (given the antioxidant and antimicrobial properties). The reuse of kiwi peel could also contribute to a circular bioeconomy and resource-use efficiency.

Supplementary Materials: The following are available online at https://www.mdpi.com/article/10.3390/app11146416/s1, Figure S1: HPLC-chromatographic profile of phenolic compounds in kiwi peel extract and chemical structure of (–)-epicatechin and quercetin-3-*O*-glucoside. The compounds identification is presented in Table 1. Figure S2: Cube plots illustrating the optimal values as a function of the three independent variables for extraction yield (A), total content of flavonoids (B), and all responses simultaneously (C). The values at each vertex of the first two cubes are response values. The model-predicted values represented in each cube plot are shown in Table 3; Table S1: Natural and codded values of the independent variables used in the five-level central composite rotatable design (CCRD) used to optimize the extraction of flavonoids from kiwi peel.; Table S2. Parametric values estimated with the polynomial Equation (1) and statistical information of the model fitting procedure. Parametric superscripted 1, 2, and 3 stand for the variables time, ultrasonic power, and ethanol concentration, respectively.

Author Contributions: Conceptualization, L.B., I.C.F.R.F., and C.C.; methodology, M.G., C.C., J.P., M.I.D., R.C.C. and D.S.; software, J.P.; validation, M.S., A.L.C., I.C.F.R.F. and L.B.; formal analysis, M.G., C.C., J.P., M.I.D., R.C.C. and D.S.; writing—original draft preparation, M.G., C.C., J.P. and M.I.D.; writing—review and editing, J.P., R.C.C., D.S., M.S., D.T., A.L.C., I.C.F.R.F. and L.B.; supervision, C.C., A.L.C., I.C.F.R.F. and L.B.; funding acquisition, L.B. All authors have read and agreed to the published version of the manuscript.

Funding: This research conducted under the project "BIOMA—Bioeconomy integrated solutions for the mobilization of the Agro-food market" (POCI-01-0247-FEDER-046112), by "BIOMA" Consortium, and financed by European Regional Development Fund (ERDF), through the Incentive System to Research and Technological development, within the Portugal2020 Competitiveness and Internationalization Operational Program. This work has been supported by the Ministry of Education, Science and Technological Development of Republic of Serbia (451-03-9/2021-14/200007).

Institutional Review Board Statement: Not applicable.

Informed Consent Statement: Not applicable.

Acknowledgments: The authors are grateful to the Foundation for Science and Technology (FCT, Portugal) for financial support by national funds FCT/MCTES to CIMO (UIDB/00690/2020). M.I. Dias, R.C. Calhelha, and L. Barros thank the national funding by FCT, P.I., through the institutional scientific employment program-contract, and J. Pinela through the individual scientific employment program-contract (CEECIND/01011/2018). C. Caleja thanks her contract through the project Healthy-PETFOOD (POCI-01-0247-FEDER-047073). The authors also thank the company KiwiCoop (Oliveira do Bairro, Portugal) for providing the kiwi samples.

Conflicts of Interest: The authors declare no conflict of interest.

References

1. Food and Agriculture Organization of the United Nations (FAO) Food Loss and Food Waste | Policy Support and Governance Gateway. Available online: http://www.fao.org/policy-support/policy-themes/food-loss-food-waste/en/ (accessed on 27 June 2021).
2. Sagar, N.A.; Pareek, S.; Sharma, S.; Yahia, E.M.; Lobo, M.G. Fruit and vegetable waste: Bioactive compounds, their extraction and possible utilization. *Compr. Rev. Food Sci. Food Saf.* **2018**, *17*, 512–531. [CrossRef] [PubMed]
3. Nanda, S.; Isen, J.; Dalai, A.K.; Kozinski, J.A. Gasification of fruit wastes and agro-food residues in supercritical water. *Energy Convers. Manag.* **2016**, *110*, 296–306. [CrossRef]
4. Mateos-Aparicio, I. Plant-based by-products. In *Food Waste Recovery: Processing Technologies, Industrial Techniques, and Applications*; Galanakis, C.M., Ed.; Elsevier: Amsterdam, The Netherlands, 2021; pp. 367–397.
5. Mahato, N.; Sharma, K.; Sinha, M.; Cho, M.H. Citrus waste derived nutra-/pharmaceuticals for health benefits: Current trends and future perspectives. *J. Funct. Foods* **2018**, *40*, 307–316. [CrossRef]
6. Lavelli, V. Circular food supply chains—Impact on value addition and safety. *Trends Food Sci. Technol.* **2021**, *114*, 323–332. [CrossRef]
7. Liu, Y.; Qi, Y.; Chen, X.; He, H.; Liu, Z.; Zhang, Z.; Ren, Y.; Ren, X. Phenolic compounds and antioxidant activity in red- and in green-fleshed kiwifruits. *Food Res. Int.* **2019**, *116*, 291–301. [CrossRef]
8. Park, Y.S.; Namiesnik, J.; Vearasilp, K.; Leontowicz, H.; Leontowicz, M.; Barasch, D.; Nemirovski, A.; Trakhtenberg, S.; Gorinstein, S. Bioactive compounds and the antioxidant capacity in new kiwi fruit cultivars. *Food Chem.* **2014**, *165*, 354–361. [CrossRef] [PubMed]
9. Sanz, V.; López-Hortas, L.; Torres, M.D.; Domínguez, H. Trends in kiwifruit and byproducts valorization. *Trends Food Sci. Technol.* **2021**, *107*, 401–414. [CrossRef]
10. Kumar, K.; Srivastav, S.; Sharanagat, V.S. Ultrasound assisted extraction (UAE) of bioactive compounds from fruit and vegetable processing by-products: A review. *Ultrason. Sonochem.* **2021**, *70*, 105325. [CrossRef]

11. Caleja, C.; Barros, L.; Prieto, M.A.; Barreiro, F.M.F.; Oliveira, M.B.P.; Ferreira, I.C.F.R. Extraction of rosmarinic acid from *Melissa officinalis* L. by heat-, microwave- and ultrasound-assisted extraction techniques: A comparative study through response surface analysis. *Sep. Purif. Technol.* **2017**, *186*, 297–308. [CrossRef]
12. Albuquerque, B.R.; Prieto, M.A.; Vazquez, J.A.; Barreiro, M.F.; Barros, L.; Ferreira, I.C F.R. Recovery of bioactive compounds from *Arbutus unedo* L. fruits: Comparative optimization study of maceration/microwave/ultrasound extraction techniques. *Food Res. Int.* **2018**, *109*, 455–471. [CrossRef]
13. Bessada, S.M.F.; Barreira, J.C.M.; Barros, L.; Ferreira, I.C.F.R.; Oliveira, M.B.P.P. Phenolic profile and antioxidant activity of *Coleostephus myconis* (L.) Rchb.f.: An underexploited and highly disseminated species. *Ind. Crops Prod.* **2016**, *89*, 45–51. [CrossRef]
14. Mandim, F.; Barros, L.; Calhelha, R.C.; Abreu, R.M.V.; Pinela, J.; Alves, M.J.; Heleno, S.; Santos, P.F.; Ferreira, I.C.F.R. *Calluna vulgaris* (L.) Hull: Chemical characterization, evaluation of its bioactive properties and effect on the vaginal microbiota. *Food Funct.* **2019**, *10*, 78–89. [CrossRef]
15. Barros, L.; Pereira, E.; Calhelha, R.C.; Dueñas, M.; Carvalho, A.M.; Santos-Buelga, C.; Ferreira, I.C.F.R. Bioactivity and chemical characterization in hydrophilic and lipophilic compounds of *Chenopodium ambrosioides* L. *J. Funct. Foods* **2013**, *5*, 1732–1740. [CrossRef]
16. Vaz, J.A.; Heleno, S.A.; Martins, A.; Almeida, G.M.; Vasconcelos, M.H.; Ferreira, I.C.F.R. Wild mushrooms *Clitocybe alexandri* and *Lepista inversa*: In vitro antioxidant activity and growth inhibition of human tumour cell lines. *Food Chem. Toxicol.* **2010**, *48*, 2881–2884. [CrossRef] [PubMed]
17. Soković, M.; Glamočlija, J.; Marin, P.D.; Brkić, D.; Griensven, L.J.L.D. van Antibacterial effects of the essential oils of commonly consumed medicinal herbs using an in vitro model. *Molecules* **2010**, *15*, 7532–7546. [CrossRef]
18. Soković, M.; van Griensven, L.J.L.D. Antimicrobial activity of essential oils and their components against the three major pathogens of the cultivated button mushroom, *Agaricus bisporus*. *Eur. J. Plant Pathol.* **2006**, *116*, 211–224. [CrossRef]
19. Albuquerque, B.R.; Prieto, M.A.; Barreiro, M.F.; Rodrigues, A.; Curran, T.P.; Barros, L.; Ferreira, I.C.F.R. Catechin-based extract optimization obtained from *Arbutus unedo* L. fruits using maceration/microwave/ultrasound extraction techniques. *Ind. Crops Prod.* **2017**, *95*, 404–415. [CrossRef]
20. Pinela, J.; Prieto, M.A.; Carvalho, A.M.; Barreiro, M.F.; Oliveira, M.B.P.P.; Barros, L.; Ferreira, I.C.F.R. Microwave-assisted extraction of phenolic acids and flavonoids and production of antioxidant ingredients from tomato: A nutraceutical-oriented optimization study. *Sep. Purif. Technol.* **2016**, *164*, 114–124. [CrossRef]
21. Anjos, R.; Cosme, F.; Gonçalves, A.; Nunes, F.M.; Vilela, A.; Pinto, T. Effect of agricultural practices, conventional vs. organic, on the phytochemical composition of 'Kweli' and 'Tulameen' raspberries (*Rubus idaeus* L.). *Food Chem* **2020**, *328*, 126833. [CrossRef]
22. Latocha, P.; Krupa, T.; Wołosiak, R.; Worobiej, E.; Wilczak, J. Antioxidant activity and chemical difference in fruit of different *Actinidia* sp. *Int. J. Food Sci. Nutr.* **2010**, *61*, 381–394. [CrossRef]
23. Rocha, R.; Pinela, J.; Abreu, R.M.V.; Añibarro-Ortega, M.; Pires, T.C.S.P.; Saldanha, A.L.; Alves, M.J.; Nogueira, A.; Ferreira, I.C.F.R.; Barros, L. Extraction of anthocyanins from red raspberry for natural food colorants development: Processes optimization and in vitro bioactivity. *Processes* **2020**, *8*, 1447. [CrossRef]
24. Iberahim, N.; Sethupathi, S.; Goh, C.L.; Bashir, M.J.K.; Ahmad, W. Optimization of activated palm oil sludge biochar preparation for sulphur dioxide adsorption. *J. Environ. Manag.* **2019**, *248*, 109302. [CrossRef]
25. Dias, M.; Caleja, C.; Pereira, C.; Calhelha, R.C.; Kostic, M.; Sokovic, M.; Tavares, D.; Baraldi, I.J.; Barros, L.; Ferreira, I.C.F.R. Chemical composition and bioactive properties of byproducts from two different kiwi varieties. *Food Res. Int.* **2020**, *127*, 108753. [CrossRef]
26. Bernardes, N.R.; Talma, S.V.; Sampaio, S.H.; Nunes, C.R.; Rangel de Almeida, J.A.; De Oliveira, D.B. Atividade antioxidante e fenóis totais de frutas de Campos dos Goytacazes RJ. *Biol. Saúde* **2011**, *1*. [CrossRef]
27. Fiorentino, A.; Mastellone, C.; D'Abrosca, B.; Pacifico, S.; Scognamiglio, M.; Cefarelli, G.; Caputo, R.; Monaco, P. δ-Tocomonoenol: A new vitamin E from kiwi (*Actinidia chinensis*) fruits. *Food Chem.* **2009**, *115*, 187–192. [CrossRef]
28. Soquetta, M.B. *Physicochemical, Microbiological Characterization and Bioactive Compounds of Kiwi (Actinidia deliciosa) Peel and Pomace Flour and Application in Pâté*; Universidade Federal de Santa Maria: Santa Maria, Brazil, 2015.
29. Moita, J.P.R. *Influência de Infusões de Plantas na Permeação de Aminoácidos Pela Barreira Intestinal e Identificação de Compostos Bioativos Com Propriedades Antioxidantes*; Universidade de Lisboa: Lisboa, Portugal, 2015.
30. Lim, S.; Han, S.H.; Kim, J.; Lee, H.J.; Lee, J.G.; Lee, E.J. Inhibition of hardy kiwifruit (*Actinidia aruguta*) ripening by 1-methylcyclopropene during cold storage and anticancer properties of the fruit extract. *Food Chem.* **2016**, *190*, 150–157. [CrossRef] [PubMed]
31. An, X.; Lee, S.G.; Kang, H.; Heo, H.J.; Cho, Y.S.; Kim, D.O. Antioxidant and anti-inflammatory effects of various cultivars of kiwi berry (*Actinidia arguta*) on lipopolysaccharide-stimulated raw 264.7 cells. *J. Microbiol. Biotechnol.* **2016**, *26*, 1367–1374. [CrossRef] [PubMed]
32. El Kichaoi, A.; El-Hindi, M.; Mosleh, F.; Elbashiti, T. The antimicrobial effects of the fruit extracts of *Punica granatum*, *Actinidia deliciosa* and *Citrus maxima* on some human pathogenic microorganisms. *Am. Int. J. Biol.* **2015**, *3*, 63–75. [CrossRef]
33. Gavrović-Jankulović, M.; Ćirković, T.; Vučković, O.; Atanasković-Marković, M.; Petersen, A.; Gojgić, G.; Burazer, L.; Jankov, R.M. Isolation and biochemical characterization of a thaumatin-like kiwi allergen. *J. Allergy Clin Immunol.* **2002**, *110*, 805–810. [CrossRef]
34. Xia, L.; Ng, T.B. Actinchinin, a novel antifungal protein from the gold kiwi fruit. *Peptides* **2004**, *25*, 1093–1098. [CrossRef]

MDPI

Article

The Concentration-Dependent Effects of Essential Oils on the Growth of *Fusarium graminearum* and Mycotoxins Biosynthesis in Wheat and Maize Grain

Daniela Gwiazdowska [1,*], Katarzyna Marchwińska [1], Krzysztof Juś [1], Pascaline Aimee Uwineza [2], Romuald Gwiazdowski [3], Agnieszka Waśkiewicz [2] and Roman Kierzek [4]

1 Department of Natural Science and Quality Assurance, Institute of Quality Science, Poznań University of Economics and Business, Niepodległości 10, 61-875 Poznań, Poland; katarzyna.marchwinska@ue.poznan.pl (K.M.); krzysztof.jus@ue.poznan.pl (K.J.)

2 Department of Chemistry, Poznań University of Life Sciences, Wojska Polskiego 75, 60-625 Poznań, Poland; pascaline.uwineza@up.poznan.pl (P.A.U.); agnieszka.waskiewicz@up.poznan.pl (A.W.)

3 Research Centre for Registration of Agrochemicals, Institute of Plant Protection-National Research Institute, Władysława Węgorka 20, 60-318 Poznań, Poland; r.gwiazdowski@iorpib.poznan.pl

4 Department of Weed Science and Plant Protection Technique, Institute of Plant Protection-National Research Institute, Władysława Węgorka 20, 60-318 Poznań, Poland; r.kierzek@iorpib.poznan.pl

* Correspondence: daniela.gwiazdowska@ue.poznan.pl; Tel.: +48-61-856-95-36

Featured Application: In these times, in accordance with the European Green Deal and the "Farm to Fork Strategy" there should be a strong emphasis on efforts to create a healthier and more sustainable food system. One of the key targets in this plan is the reduction by 2030 of the overall use of chemical pesticides by 50%, which prompts scientists to intensify research in the search for alternative solutions. Healthy food means safe food, free from pathogens and toxic metabolites, such as *Fusarium* fungi and their mycotoxins. Essential oils are a noteworthy alternative, since more and more studies have proven their strong antifungal and antimycotoxigenic activity. The presented research broadens and complements the existing knowledge in the field of practical application of oils in biological control of plant pathogens.

Abstract: The presence of *Fusarium* fungi and their toxic metabolites in agricultural crops contributes to significant quantitative and qualitative losses of crops, causing a direct threat to human and animal health and life. Modern strategies for reducing the level of fungi and mycotoxins in the food chain tend to rely on natural methods, including plant substances. Essential oils (EOs), due to their complex chemical composition, show high biological activity, including fungistatic properties, which means that they exhibit high potential as a biological plant protection factor. The aim of this study was to determine the fungistatic activity of three EOs against *F. graminearum*, and the reduction of mycotoxin biosynthesis in corn and wheat grain. All tested EOs effectively suppressed the growth of *F. graminearum* in concentrations of 5% and 10%. Cinnamon and verbena EOs also effectively reduced the ergosterol (ERG) content in both grains at the concentration of 1%, while at the 0.1% EO concentration, the reduction in the ERG amount depended on the EO type as well as on the grain. The degree of zearalenone (ZEA) reduction was consistent with the inhibition of ERG biosynthesis, while the reduction in deoxynivalenol (DON) was not consistent with this parameter.

Keywords: antifungal activity; biological plant protection; cereals quality; essential oils; ergosterol; food chain safety; *Fusarium* spp.; mycotoxins

Citation: Gwiazdowska, D.; Marchwińska, K.; Juś, K.; Uwineza, P.A.; Gwiazdowski, R.; Waśkiewicz, A.; Kierzek, R. The Concentration-Dependent Effects of Essential Oils on the Growth of *Fusarium graminearum* and Mycotoxins Biosynthesis in Wheat and Maize Grain. *Appl. Sci.* **2022**, *12*, 473. https://doi.org/10.3390/app12010473

Academic Editor: Ramona Iseppi

Received: 19 October 2021
Accepted: 30 December 2021
Published: 4 January 2022

1. Introduction

Wheat (*Triticum aestivum* L.) and maize (*Zea mays* L.) are the most important cereal crops worldwide, with over 775.8 million and 1125 billion metric tons produced in 2020/21, respectively [1]. Both cereals are used as food ingredients and animal feed, as they are good sources of nutrients. Moreover, different types of food and industrial products, including sweeteners, starch, oil, beverages, glue, industrial alcohol, and fuel ethanol, can be produced from wheat and maize. The consumption of these cereals varies mainly by region; maize is consumed highly in Central America, Southern and Eastern Africa, and Mexico, while wheat is preferred mainly in Central Asia, the Middle East, South and North America, and Europe [2]. A serious problem in wheat and corn crops consists of infections caused by the *Fusarium* genus, the filamentous fungi commonly found in soil and on plants, contaminating agricultural commodities both on the field and in storage [3]. *Fusarium graminearum* is one of the most prevalent species that infect cereals in humid and semi-humid areas worldwide [4], responsible for devastating and hazardous plant diseases, such as stalk and ear rot of maize, as well as *Fusarium* head blight (FHB), crown rot, and seedling blight on wheat and other cereals [5]. *Fusarium* diseases cause losses not only due to yield, quality and nutritional properties reduction, but also because contamination of grains with mycotoxins affects food and feed safety, posing a threat to human and animals' health [4]. The most widespread mycotoxins associated with *F. graminearum* infection in cereals include zearalenone (ZEA) and trichothecenes, such as nivalenol (NIV), deoxynivalenol (DON), and its acetylated derivatives 3-acetyl-deoxynivalenol (3-ADON), 15-acetyl-deoxynivalenol (15-ADON) and deoxynivalenol 3 β–D-glucoside (DON-3-G) [6–8]. ZEA is considered an estrogenic mycotoxin and found to be toxic for the liver, kidney, and immune system [9], while trichothecenes can inhibit protein synthesis and induce anorexia, diarrhea, vomiting, and cell death, along with altering immune function, phosphokinase-mediated stress pathways, proinflammatory gene expression, gastrointestinal function, and the action of growth hormones [6,10].

Up to now, synthetic fungicides have been the primary tool to control contamination by *Fusarium* fungi and their mycotoxins, while many other policies have been developed and implemented to inhibit or reduce fungal growth and mycotoxin biosynthesis in different crops, notably in cereals. In recent years, the use of chemicals has increased consumer awareness, and their use is becoming more challenging due to carcinogenic effects, residual toxicity problems, environmental pollution, and increasing microbial resistance [11]. Therefore, finding alternatives to chemical fungicides that are considered safe, with negligible risk to human health and the environment, has become an exciting topic among researchers. EOs, which are naturally antimicrobial and antifungal substances, have received much attention and are considered as the biological antifungal agents that can replace synthetic pesticides [4,12]. Their significant benefit is that many EOs are on the GRAS (Generally Recognized as Safe) list approved by the FDA (Food and Drug Administration), therefore they can be stated as safe for the consumers and the environment [13].

EOs are natural secondary metabolites, consisting mainly of monoterpenes and sesquiterpenes [14]. EOs are accumulated in cells, secretory cavities, or glandular hairs of plants, and they can be derived from one or more plant parts, such as the root (e.g., vetiver grass, angelica, valerian), stems and leaves (e.g., verbena, petitgrain, patchouli, geranium), leaves (e.g., mint, lemongrass, lemon balm, citronella), buds (e.g., clove), flowers (e.g., lavender, rose, jasmine, carnation, clove, mimosa, rosemary), fruits (e.g., lemon, orange, juniper) or their peels (e.g., orange, lemon, grapefruit), bark (e.g., cinnamon), rhizomes (e.g., ginger, calamus, curcuma, orris), and seeds (e.g., fennel, coriander, caraway, dill, nutmeg, pepper) [14,15]. There are different methods for EO extraction, such as hydrodistillation, steam distillation, maceration, CO_2 extraction, solvent extraction, enfleurage, and cold press extraction [15]. Furthermore, different studies have shown that the extraction method has an influence on the content of individual compounds, the yield of EOs, and their activities. Due to different constituents of EOs and their biological properties (antioxidant, antibacterial, antifungal), they have wide applications, among others, in cosmetics (e.g., fragrances and lotions), in food products (e.g., flavorings and preservatives), in agriculture (e.g., insecticides and repellents), and in pharmaceutical products (e.g., therapeutic action and sanitary activity) [16–19].

Studies conducted in recent years show that the application of EOs to control the pathogenic *Fusarium* spp. that attack cereals and cereal products is very effective. For instance, mycelial growth was significantly or completely inhibited in the tested isolates of *Fusarium* (*F. oxysporum*, *F. solani*, *F. verticillioides*, and *F. subglutinans*) by EOs of eucalyptus, clove, lemongrass, and mint [20,21]. In turn, the EOs of bay leaf, cinnamon, clove and oregano reduced fungal growth of *F. culmorum* and *F. verticillioides* by 90% to almost 100% [22]. In research conducted by Kalagatur et al. [23], the use of *Cymbopogon martinii* EO (from leaves of tropical herbaceous grass) in the encapsulated form with chitosan has significantly inhibited the growth of *F. graminearum* through the reduction of ergosterol (ERG) content and increase in intracellular reactive oxygen species and lipid peroxidation [23]. In a similar study, Ferreira and co-workers have reported the powerful effects of ginger EO against *F. graminearum* in vitro through the ERG and DON reduction [4]. The inhibitory effect of EOs showed dose-dependent activity on the tested *Fusarium* spp.

In the presented study, the antifungal and anti-mycotoxigenic effects of various concentrations of commercial palmarosa, cinnamon, and verbena EOs on the growth of *F. graminearum* and biosynthesis of mycotoxins in wheat and maize grain were analyzed. The aim of the present study was to determine an inhibitory dose that effectively inhibits both the growth of *F. graminearum* and mycotoxin biosynthesis. To our best knowledge, there are still limited data of complex EO analyses taking into account the relationship between the concentration and the effectiveness of their action. In relation to earlier research, where the main purpose was comparison of the antifungal and antimycotoxigenic activity of selected EOs against *F. graminearum* and *F. culmorum* in wheat and maize grains, a wider range of experimental concentrations was used, as well as extended scope of research including advanced microscopic analyses. It is also worth underlining that the issues raised in this work are especially important considering the potential application of EOs.

2. Materials and Methods

2.1. Plant Material

Wheat (cultivar Jantarka) and maize (cultivar Wiarus) grain for the research was obtained from the Research Centre for Registration of Agrochemicals, Institute of Plant Protection, National Research Institute, in Poznań, Poland. The chemical composition of wheat grains was as follows: protein 11.2% DM, starch 67.5% DM, gluten 18.7%, moisture after storage 11.5%, and thousand grain weight (TGW) 33.7 g, while the composition of maize grains was as follows: protein 12.5 % DM, starch 68.4 % DM, oil 4.6%, moisture after storage 13%, and thousand grain weight (TGW) 312.5 g. Analyses were conducted with the FOSS Infratec™ 1241 Grain Analyser. The samples (50 g) were mixed with 10 mL of deionized water and sterilized at 121 °C.

2.2. Fusarium Strain

F. graminearum, KZF-1 was obtained from the collection of the Research Centre for Registration of Agrochemicals, Institute of Plant Protection, National Research Institute, in Poznań, Poland. The tested strain was cultured before the experiment in Petri dishes (9 cm diameter) on a PDA medium (Potato Dextrose Agar, A&A Biotechnology) at 25 °C for 5–7 days.

2.3. Standards, Chemicals, and Reagents

ZEA, DON, and ERG analytical standards were purchased from Sigma-Aldrich (Steinheim, Germany). LC-MS acetonitrile, methanol, and water (MS grade) were purchased from J.T. Baker (Deventer, The Netherlands). Chemical reagents necessary for the extraction and purification process were obtained from Sigma-Aldrich and POCh (Gliwice, Poland). Individual stock solutions for all analytes were prepared by dissolving in acetonitrile.

2.4. EOs Preparations

Studies included selected EOs, such as cinnamon bark (*Cinnamomum zeylanicum*, Indonesia), and palmarosa leaves (*Cymbopogon martini*, India), as well as verbena leaves and flowers (*Thymus hiemalis*, Spain). Composition of tested EOs was as following: (i) cinnamon bark: cinnamic aldehyde $\leq$ 70%, eugenol $\leq$ 4.4%, linalool $\leq$ 2.6%, limonene $\leq$ 1.1%, benzyl benzoate $\leq$ 1.1%, benzaldehyde 0.5%, cinnamic alcohol $\leq$ 0.4%, and cuminaldehyde $\leq$ 0.2%; (ii) palmarosa: geraniol 85%, linalool 2–3%, limonene 1%, and citral 1%; (iii) verbena: citral 42%, and limonene 40%. Commercial EOs were purchased from Ecospa s.c., Poland. The studies included four different concentrations of EO deionized water solutions (10%, 5%, 1% and 0.1%) containing 10% Tween 80.

2.5. Antifungal Activity of EOs by Disc Volatilization Method

The antifungal activity of EOs was determined by the disc volatilization method according to [24,25], with some modifications. The Petri plates, containing 10 mL of the PDA medium, were inoculated with 6 mm diameter plugs of the actively growing *F. graminearum* mycelium from 5-day old cultures. Next, 6 mm diameter sterile paper discs (Oxoid, Basingstoke, Hants, United Kingdom) were placed on the inner surface of the Petri dishes lids (one on each) and soaked with 5 µL of four different concentrations of the EOs. A sample with a water-soaked blank paper disc was used as a control. The Petri plates were immediately closed, sealed with parafilm and incubated until the mycelium in the control reached the edge of the plate. After incubation, the diameter of the mycelium was measured, and the antifungal activity was expressed as the percentage inhibition of mycelial growth in relation to the control.

2.6. EOs and the Growth of Fusarium and Mycotoxins Biosynthesis

EO solution samples in the amount of 5 mL were mixed thoroughly with 50 g of sterilized wheat and maize grains (depending on the samples) in Erlenmeyer flasks, under sterile conditions. Three plugs (6 mm) of 5-day old fungal mycelium grown on PDA were added to the maize grains. For a given series of samples/replicates, the mycelium plugs were cut out from the same plate at an equal distance from the edges of the colony to ensure the same state of development of the fungal mycelium. Next, the samples were incubated in a dark room at 25 °C for 28 days. After incubation, the samples were dried, milled, homogenized, and prepared for chromatographic analysis. Control samples were grains with the addition of deionized water and Tween 80 (without the addition of EOs).

2.7. Chemical Analysis

2.7.1. Ergosterol

ERG is a typical component of fungal cell membranes that is not present in higher plants; therefore, it can be used as a natural, selective indicator for the presence of fungi in different matrices. Dried and milled samples of maize or wheat grain in the amount of 100 mg were suspended in 2 mL of methanol and 0.5 mL of a 2 M aqueous solution of sodium hydroxide. Next, samples in tubes were tightly screwed and microwaved three times (370 W power) for 10 s, and stored at room temperature for cooling down. Subsequently, samples were neutralized with 1 mL of 1 M hydrochloric acid solution, and 2 mL of methanol. After mixing, triple extraction by 4 mL of n-pentane was performed. Each time, the n-pentane layer was transferred to the vials and dried under a stream of nitrogen. The dry residue was dissolved in 1 mL of methanol and filtered through a 0.20 µm syringe filter (Chromafil, Macherey-Nagel, Duren, Germany) before HPLC/PDA analysis. ERG was detected on a HPLC Waters Alliance system with a Waters 2996 Photodiode Array Detector (Waters Division of Millipore, Milford, MA, USA) set at 282 nm and a 3.9 × 150 mm Nova Pak C-18 chromatographic column. The mobile phase was methanol:acetonitrile (90:10, *v*/*v*) at a flow rate of 1.0 mL/min. The concentration of ERG was determined as a result of the comparison of retention times with the external standard. The detection limit was 10 ng/g.

2.7.2. Mycotoxins Analysis

The extraction of mycotoxins was carried out by adding up to 5 g of ground samples to 20 mL of the extraction mixture (acetonitrile:water:acetic acid, 79:20:1, *v*/*v*/*v*), followed by vortexing (about 30 s) and mixing using a horizontal shaker for 24 h. After extraction, the samples were centrifuged at 3000 rpm for 10 min, and finally, all samples were filtered through a 0.20 µm syringe filter (Chromafil, Macherey-Nagel, Duren, Germany) before LC/MS/MS analyses.

The analytical system consisted of the Aquity UPLC chromatography (Waters, Manchester, MA, USA) coupled to an electrospray ionization triple quadrupole mass spectrometer (TQD) (Waters, Manchester, MA, USA). A Waters ACQUITY UPLC HSS T3 (100 × 2.1 mm/ID, with 1.8 µm particle size) (Waters, Manchester, MA, USA) was used for chromatographic separation with a flow rate 0.35 mL/min at room temperature. Gradient elution was applied using water buffered with 10 mM ammonium acetate (A) and acetonitrile (B). The solvent gradient was modified as follows: 0–2 min at 5% B and 2–7 min at 55% B. Next, these conditions were maintained for 2 min and 9–15 min at 90% B with isocratic elution for 2 min, followed by a return to the initial conditions. Nitrogen above 99% purity was used. The collision-induced decomposition was run using argon as the collision gas, with a collision energy of 14–22 eV. Multiple reaction monitoring (MRM) was used for quantitative analysis of the compounds. The analytes were identified by comparing the retention times and m/z values obtained by MS and MS^2 with the mass spectra (317.1/174.9 and 297.3/249.1 for ZEA and DON, respectively) of the corresponding standards tested under the same conditions. All samples were injected in triplicate.

2.8. Morphology Observations

F. graminearum was cultured in PDB (Potato Dextrose Broth) for 5 days at 25 °C. Then cultures were centrifugated and treated with different concentrations of EOs for 1 h at 25 °C. Samples were transferred on glass slides and photographed with a light microscope (Olympus BX53, Olympus Corporation, Tokyo, Japan) at 400 × magnification. Fungal samples treated with sterile water were used as control samples. Each assay was repeated in triplicate.

2.9. Staining and Fluorescence Microscopy

The fluorescent probes, carboxyfluorescein diacetate (cFDA), and propidium iodide (PI) (Sigma Aldrich, Steinheim, Germany) were applied to assess the viability of *F. graminearum*. The procedure of staining was based on the method described by [26,27] with some modifications. A stock solution of cFDA (1 mg/mL) was prepared in dimethyl sulfoxide (DMSO), and a stock solution of PI (1 mg/mL) was prepared in distilled water and stored in the refrigerator. Centrifuged cultures of *F. graminearum*, prepared as described in Section 2.8, were suspended in 5 mL of 50 mM phosphate buffer with 25 µL cFDA and 25 µL PI at 30 °C for 15 min. Next, the staining solution was removed by centrifugation (8000 rpm, 1 min). The pellet was examined under the fluorescence microscope Olympus BX53 equipped with specific wavelength filters filter set (CFDA excitation/emission: 485/530; PI excitation/emission: 538/617).

2.10. Statistical Analysis

The results are presented as the mean (±standard deviation) of three parallel replicates. The effect of EOs on the reduction of ERG, ZEA, and DON was estimated by one-way analysis of variance (ANOVA). The homogeneity of variance was tested by Levene's test and based on the results for homogeneous samples the Tukey's test, and for nonhomogeneous samples, the Games–Howell test with a p-value < 0.05 was applied. Analyses were conducted using the IBM SPSS Statistics program.

3. Results

3.1. The Inhibition of the Growth of F. graminearum Depending on the EO Concentration

The EOs used in the presented work were selected on the basis of the earlier studies [28,29]. Four different concentrations of cinnamon, palmarosa and verbena EOs were used. The results of antifungal activity towards *F. graminearum* obtained by the disc volatilization method are presented in Table 1 and Figure 1. Cinnamon and verbena EOs demonstrated strong antifungal activity at concentrations of 10% and 5%, reaching complete growth inhibition in the highest used dose, while at a concentration of 5%, the strongest activity was shown by cinnamon EO. Palmarosa EO at the highest concentration showed only a weak impact, up to 5%, however no statistical differences compared to the control were observed. At the concentrations of 1% and 0.1%, none of the used EOs showed statistically significant fungistatic activity against *F. graminearum*.

Table 1. Volatile effect of EOs on the mycelial growth of *F. graminearum*.

EO	Effects of Different EO Concentrations on Mycelium Growth							
	0.1%		1.0%		5.0%		10.0%	
	Mycelium Size [mm]	Growth Reduction [%]	Mycelium Size [mm]	Growth Reduction [%]	Mycelium Size [mm]	Growth Reduction [%]	Mycelium Size [mm]	Growth Reduction [%]
Cinnamon	50.00 ± 0.00	0	50.00 ± 0.00	0	10.50 ± 6.36	90	6.00 ± 0.00	100
Verbena	50.00 ± 0.00	0	48.00 ± 1.41	5	20.00 ± 7.07	68	6.00 ± 0.00	100
Palmarosa	50.00 ± 0.00	0	50.00 ± 0.00	0	49.75 ± 0.35	1	47.75 ± 1.77	5

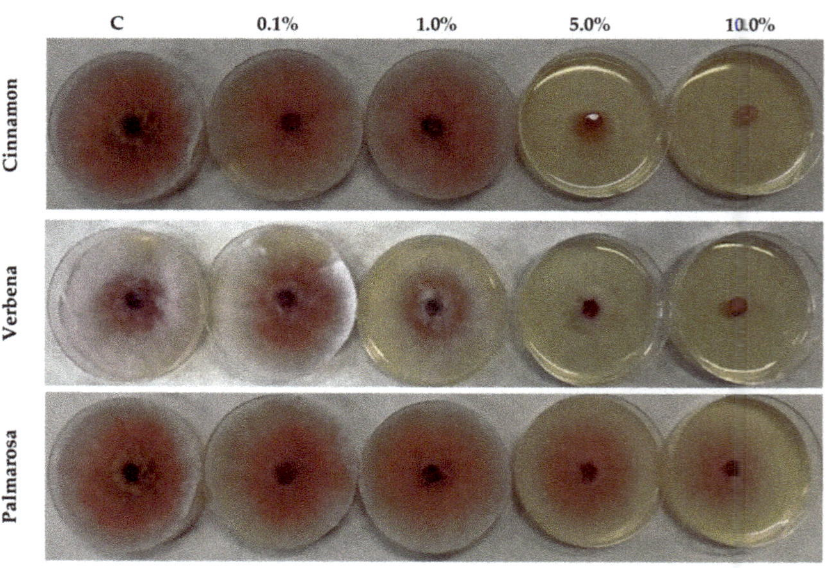

Figure 1. Volatile effects of different concentrations of palmarosa, verbena and cinnamon EOs on the mycelial growth of *F. graminearum* on PDA medium (C–control; EO tested concentrations: 0.1%; 1.0%; 5.0%; 10.0%).

3.2. Staining and Microscopy

The microscopy observations revealed the differences in the morphological appearance of hyphae. Figure 2 presents the morphology of fungal cells after incubation with cinnamon EO. In the untreated sample, hyphae were gentle and rather thin, as well as having homogeneous cell structure (Figure 2a), while in samples treated with EOs, hyphae became thicker and the internal structure was heterogeneous (Figure 2b–e). With the increasing concentrations of EOs, the changes became more visible. While in hyphae treated with 0.1% and 1% EO solutions, the heterogeneity of the internal structure was noticeable in the hyphae fragments, fungal cells treated with higher EO concentrations were completely changed. Similar observations were made for the morphology of *F. graminearum* hyphae treated with verbena EO (Figure S1 in Supplementary Material). Under the light microscope (magnification × 400), changes in the morphology of *F. graminearum* hyphae) treated with palmarosa EO were not observed, due to exhibiting the weakest activity (up to 5% growth reduction for an EO concentration of 10%).

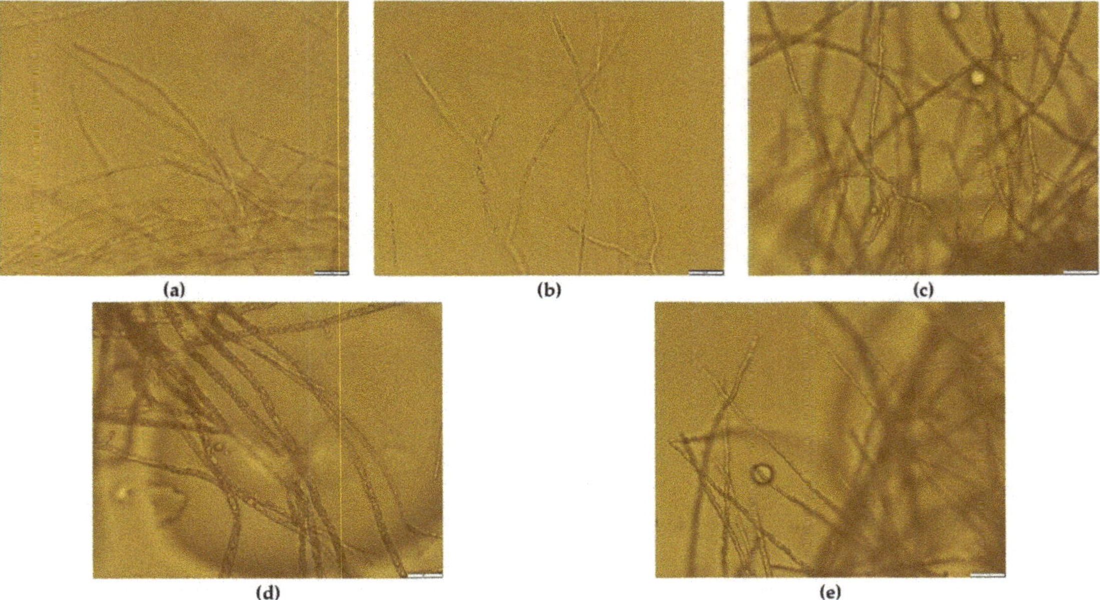

Figure 2. Morphology of *F. graminearum* hyphae under light microscope (magnification × 400) treated with cinnamon EO: (**a**) Control; (**b**) EO concentration 0.1%; (**c**) EO concentration 1.0%; (**d**) EO concentration 5.0%; (**e**) EO concentration 10.0%.

The fluorescence method using cFDA and PI was employed to evaluate the viability of fungal cells. The results of cFDA/ PI staining of *F. greminearum* treated with cinnamon EO are shown in Figure 3. cFDA is a probe that can pass through the membranes of cells [30]. In living cells, it is hydrolysed by non-specific esterases to release the fluorescent carboxyfluorescein (green fluorescence). PI is a dye that can only pass through the membranes of damaged or dead cells due to the permeability of the plasma membrane and releases red fluorescence [31].

Great differences were observed between samples treated and untreated with tested EOs. Untreated, control hyphae (Figure 3a,b) show strong green fluorescence demonstrating high viability, with only some single cells stained in red. In the samples treated with EOs, the viability of fungal cells depended on the EO concentration, as shown in the example of cinnamon oil (Figure 3c–i). In samples with 0.1%, high viability of cells was still observed, as the green fluorescence was visible and very few cells demonstrated red fluorescence. However, a higher concentration of EO caused a drastic decrease in viable cells. The fungal cells observed under the microscope showed strong red fluorescence, while green fluorescence was found. Similar findings were observed for the morphology of *F. graminearum* treated with all tested EOs (Figures S2 and S3 in Supplementary Material).

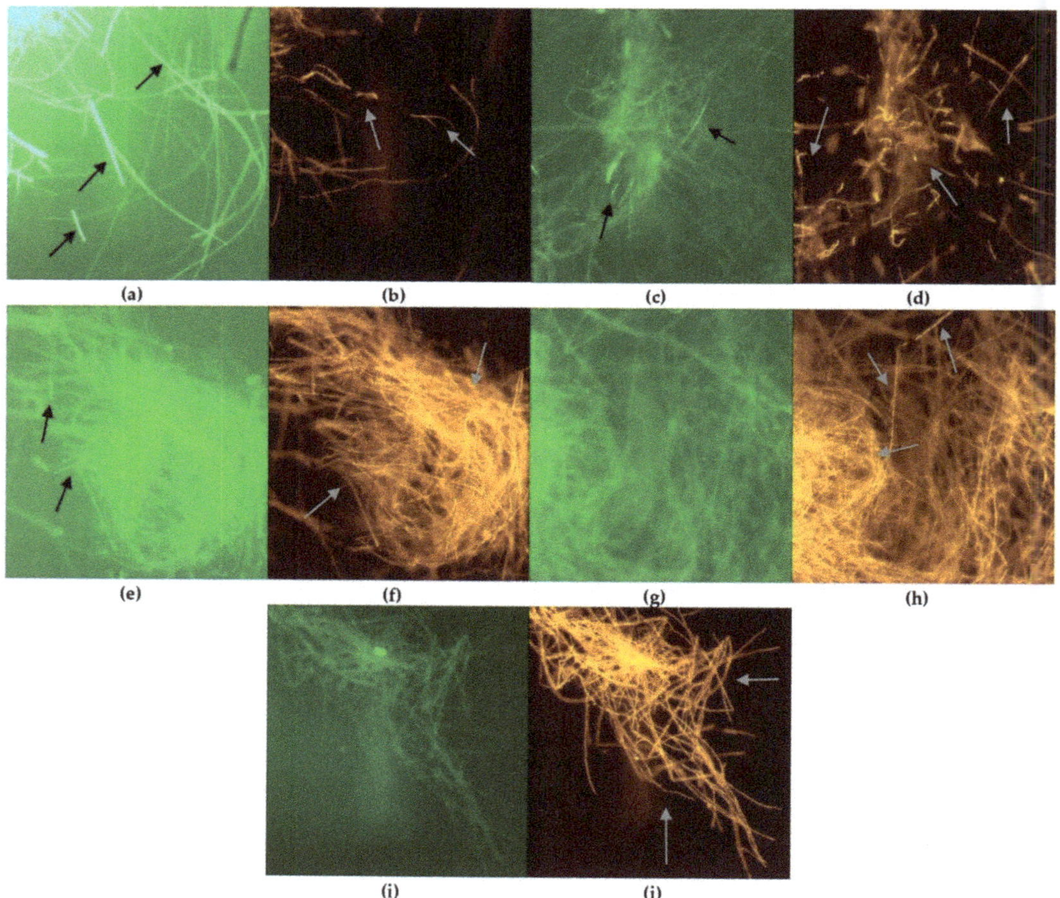

Figure 3. Morphology of *F. graminearum* hyphae under optical microscope (magnification × 400): untreated: (**a**) control, stained with cFDA; (**b**) control, stained with PI; treated with cinnamon EO: stained with cFDA, EO concentration: (**c**) 0.1%; (**e**) 1.0%; (**g**) 5.0%; (**i**) 10.0%; stained with PI, EO concentration: (**d**) 0.1%; (**f**) 1.0%; (**h**) 5.0%; (**j**) 10.0%.

3.3. The Effect of EO Concentration on the Growth of F. graminearum on Maize and Wheat Grain

The effect of EOs on the *F. graminearum* growth on maize and wheat grains was determined by the measurement of ERG concentration by HPLC analysis (Figure 4). The reduction percentage of ERG in grain samples treated with EO preparations was calculated in comparison to the control samples, where the grains were treated with a mixture of deionized water and Tween 80 without the addition of EO. The ERG was reduced significantly in the concentration range of 1% to 10% by all tested EOs, in both maize and wheat grain. Cinnamon and verbena EOs effectively reduced the ERG content in both wheat and maize grains at the concentration range of 1% to 10% (ERG level reduction in the range of 95–100%, depending on the EO and grain type). Palmarosa EO effectively inhibited the growth of *F. graminearum* at concentrations of 5% and 10%, with the ERG reduction at a level of 96–100%. In the 0.1% EO concentration, the reduction in ERG amount was strongly differentiated, depending on both the EO type and the cereal grain. In wheat grain, ERG was reduced at levels of 55–65%, depending on the EO type, while in the maize grain, only cinnamon EO exhibited significant fungistatic activity compared to the control sample.

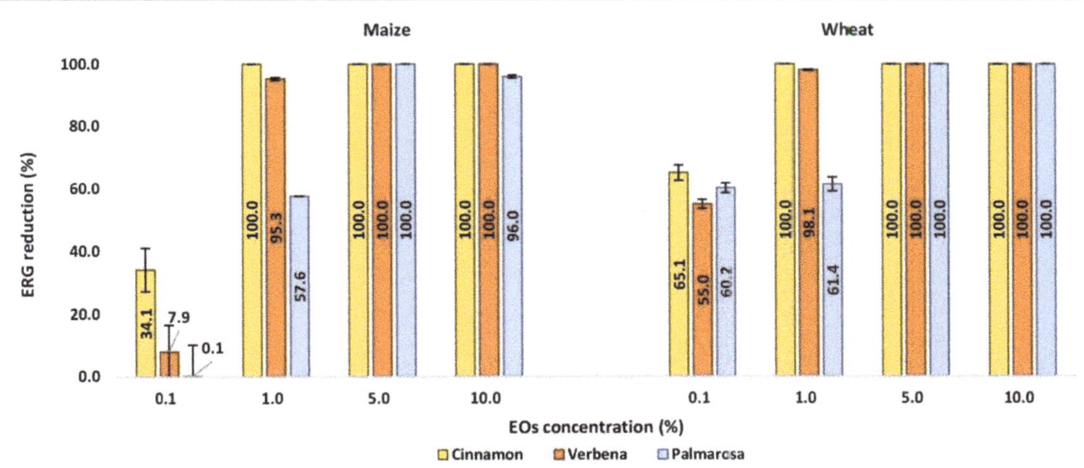

EO	Ergosterol content (µg/g)									
	Maize grain					Wheat grain				
	Control	EO concentration				Control	EO concentration			
		0.1%	1.0%	5.0%	10.0%		0.1%	1.0%	5.0%	10.0%
Cinnamon	23.20^{d} ± 2.28	15.29^{c} ±1.60	nd	nd	nd	54.97^{d} ±5.05	19.18^{b} ±1.33	nd	nd	nd
Verbena		21.38^{d} ±1.99	1.09^{a} ±0.11	nd	nd		24.74^{c} ±0.79	1.03^{a} ±0.09	nd	nd
Palmarosa		23.17^{d} ±2.32	9.83^{b} ±0.02	nd	0.93^{a} ±0.11		21.87^{bc} ±0.86	21.22^{bc} ±1.25	nd	nd

Figure 4. Effects of EOs on ERG contamination in maize and wheat samples after inoculation with the *F. graminearum*. nd-not detected, below the limit of quantification (<0.01 µg/g). Averages with different letters (a–d) for each cereal are significantly different at the $p < 0.05$.

3.4. The Effect of EO Concentration on the ZEA Level in Maize and Wheat Grain

The reduction of ZEA concentrations in cereals treated with EOs in general depended on EO concentration (Figure 5). The greatest degree of ZEA content reduction by all tested EOs, both in maize and wheat grains, was observed at EO concentrations from 5% to 10%. The percentage of ZEA reduction was close to or exceeded 99%, depending on the EO. All tested EOs significantly decreased ZEA content at a concentration of 1%, however, cinnamon and verbena EOs were more effective, with ZEA reduction at levels of 96% to 100%, depending on the EO and grain type. The percentage values of ZEA reduction by palmarosa EO were 84.6% and 66.2% in maize and wheat grain, respectively. At the concentration of 0.1%, only cinnamon and verbena EOs significantly reduced ZEA in cereal samples, with the highest efficacy being cinnamon EO in maize (62.2% ZEA content reduction) and verbena EO in wheat (88.3% ZEA reduction). Based on the obtained result, it should be noticed that ZEA concentrations in samples treated with palmarosa EO at a concentration of 0.1% was higher compared to the control, however, only in wheat grain was the difference significant.

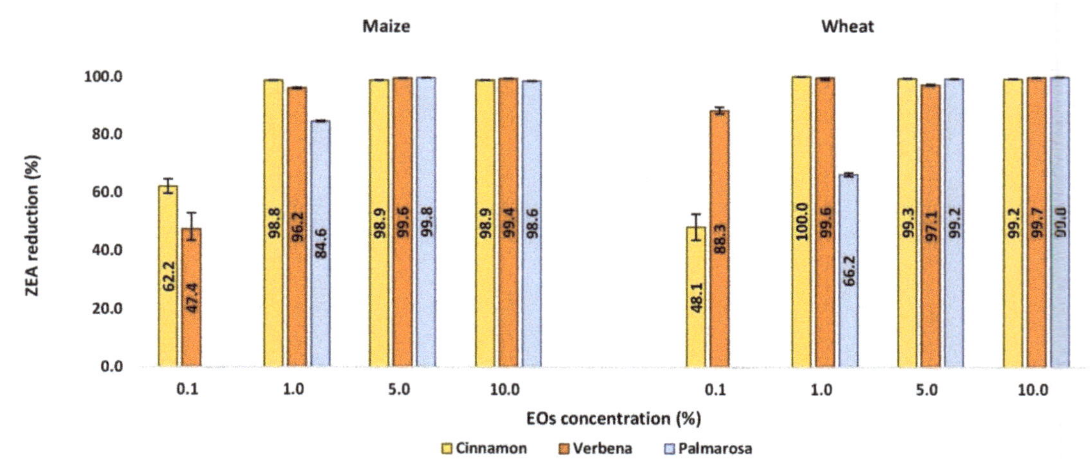

EO	Zearalenone content (μg/g)									
	Maize grain					Wheat grain				
	Control	EO concentration				Control	EO concentration			
		0.1%	1.0%	5.0%	10.0%		0.1%	1.0%	5.0%	10.0%
Cinnamon	24.85[g] ±1.72	9.39[f] ±0.61	0.30[bc] ±0.02	0.28[bc] ±0.03	0.27[b] ±0.01	47.02[h] ±4.44	24.38[g] ±2.11	nd	0.31[cd] ±0.03	0.39[d] ±0.02
Verbena		13.06[f] ±1.37	0.95[d] ±0.05	0.11[a] ±0.01	0.15[a] ±0.02		5.49[f] ±0.58	0.18[bc] ±0.01	1.38[e] ±0.15	0.16[ab] ±0.02
Palmarosa		29.24[g] ±0.98	3.82[e] ±0.05	0.06[a] ±0.01	0.34[c] ±0.01		84.27[i] ±0.47	15.88[g] ±0.30	0.37[cd] ±0.04	0.10[a] ±0.01

Figure 5. Effects of EOs on ZEA contamination in maize and wheat samples after inoculation with the *F. graminearum*. nd-not detected, below the limit of quantification (<0.01 μg/g). Averages with different letters (a–i) for each cereal are significantly different at the $p < 0.05$.

3.5. The Effect of EO Concentration on the DON Level in Maize and Wheat Grain

DON contamination in cereal samples also depended on the concentration of EOs, however, the types of cereals and EOs likewise influenced the level of toxin reduction (Figure 6). It is worth underlining that *F. graminearum* produced a low amount of DON compared to ZEA, both in maize and wheat grain (the level of DON concentration in control samples reached 3.49 and 3.19 μg/g, respectively). In wheat grain samples impregnated with tested EOs in concentrations from 1% to 10%, DON was not detected (100% of toxin reduction was noticed compared to control). At the concentration of 0.1%, no significant reduction in DON content was observed, while in samples treated with cinnamon and palmarosa EOs, a significantly higher content of the toxin was observed. In maize grain, the highest efficacy in DON reduction was demonstrated by cinnamon EO, decreasing toxin content in the whole tested range of concentration. The highest percentage reduction of the toxin was observed at the range of concentration 1% to 5% with lower efficacy at the concentration 10% (72.4% of DON reduction). Verbena and palmarosa EOs effectively reduced DON content at a concentration of 10%, while at lower concentrations, their effectiveness was much weaker. It is worth underlining that at a concentration of 0.1%, only cinnamon EO showed a significant reduction, while in samples treated with verbena and palmarosa, significantly higher content was observed.

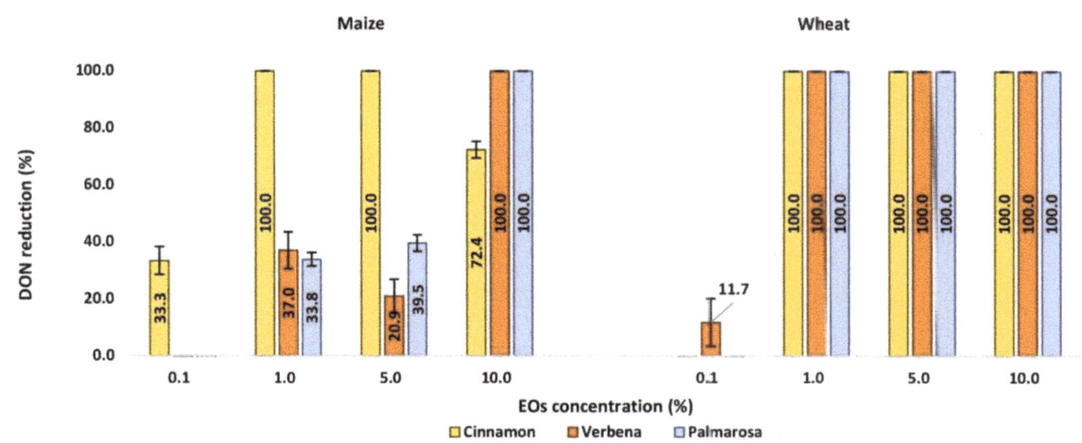

EO	Deoxynivalenol content (μg/g)									
	Maize grain					Wheat grain				
	Control	EO concentration				Control	EO concentration			
		0.1%	1.0%	5.0%	10.0%		0.1%	1.0%	5.0%	10.0%
Cinnamon	3.49^{c} ±0.21	2.33^{b} ±0.17	nd	nd	0.97^{a} ±0.10	3.19^{a} ±0.34	11.34^{c} ±0.47	nd	nd	nd
Verbena		4.37^{d} ±0.17	2.20^{b} ±0.22	2.76^{bc} ±0.21	nd		2.81^{a} ±0.27	nd	nd	nd
Palmarosa		8.20^{e} ±0.76	2.31^{b} ±0.08	2.11^{b} ±0.10	nd		6.04^{b} ±0.60	nd	nd	nd

Figure 6. Effects of EOs on DON contamination in maize and wheat samples after inoculation with the *F. graminearum*. nd-not detected, below the limit of quantification (<0.01 μg/g). Averages with different letters (a–e) for each cereal are significantly different at the $p < 0.05$.

4. Discussion

The occurrence of filamentous fungi and their toxic metabolites in food and feed has become one of the biggest global problems [32]. *Fusarium* fungi and mycotoxins constitute a significant negative factor in food production, due to the infestation of economically important cereal crops and adverse effects on human and animal health [33]. Therefore, providing a suitable amount of good quality and safe food with respect to environmental conditions is one of the priority goals in modern food production, especially at the agricultural stage [34]. There are many concepts focused on the issue of sustainable, eco-friendly food production, such as sustainable agriculture development [35], One Health concept [36] or Climate-smart agriculture (CSA) [34]. Among various methods currently used to reduce the presence of fungi and mycotoxins in food, such as physical, chemical, biological methods or good agrotechnical practices [37,38], more emphasis is placed on biological methods, especially in the context of limiting the use of pesticides in agriculture production [39]. Many data indicate the high potential of natural substances of plant origin as biological control agents, such as EOs, which are regarded as efficient, safe, environmentally friendly factors [40]. In this research, three EOs in different concentrations were tested as biological control agents against *Fusarium graminearum* growth and mycotoxin synthesis in maize and wheat grain. The presented work shows a different approach to estimating the influence of EOs on fungal growth under in vitro conditions: volatilization influence on a Petri dish and chromatographic analyses of ERG in cereal grains, as well as level of DON and ZEA reduction in maize and wheat grain using the HPLC method, were estimated. Additionally, the effects of EOs on fungal hyphae were investigated using fluorescent microscope analysis.

The conducted research showed that the EOs used exhibit antagonistic activity against *F. graminearum* and also reduce levels of DON and ZEA in wheat and maize grain. Better results were observed for the higher concentrations of EOs (10% and 5%), and cinnamon and verbena EOs were more effective than palmarosa. Antifungal activity was strongly dependent on the EO composition. Cinnamon (*Cinnamomum zeylanicum*) is a spice collected from the bark of several trees from the genus *Cinnamomum* and *Lauraceae* family. The oil from cinnamon bark oil is rich in eugenol, cinnamaldehyde, and linalool. It is also rich in β-caryophyllene and other terpenes. The antifungal properties of cinnamon EOs have been reported as effective against pathogenic *Fusarium culmorum* and *Fusarium verticillioides* [22], *Aspergillus flavus* and *Fusarium moniliforme* [41], and *Villosiclava virens*. Additionally, cinnamon water filtrate has been tested as an antifungal *Botrytis cinerea* agent [42], and cinnamon oil nanoemulsion has been reported to have an inhibitory effect on the mycelial growth of *Aspergillus niger* [43]. Lemon verbena is a perennial flowering plant of the verbena family (*Verbenaceae*). Its EOs have been reported to have antimicrobial and antioxidant properties. The most constituent compounds of verbena are geranial, neral, limonene, 1,8-cineole, spathulenol, nerol, geraniol, *trans*-β-caryophyllene, and geranyl acetate. According to researchers' findings, these components act as fumigants, insect repellents or insecticides [44]. Moreover, the antimicrobial properties of verbena EOs have been reported as effective against the pathogens *Shigella dysenteriae, Salmonella typhimurium, Pseudomonas aeruginosa* [45], *F. avenaceum, F. culmorum, F. graminearum,* and *F. oxysporum* [46]. The palmarosa EO has an antifungal activity well documented in the literature, and its biological activity has been studied in the last years. The main constituents of palmarosa EO are geraniol (82%), geranyl acetate (9%), linalool (2%), and *trans*-β-ocimene (1%), whereas geraniol is the main constituent associated with its antimicrobial activity [47]. The use of palmarosa EO presents promising results in the control of phytopathogens [23]. It is worth emphasizing that growth inhibition of *F. graminearum* was greater in wheat and maize grain, especially in EO concentrations of 1% and 0.1%, compared to the volatilization effect on Petri dishes.

Moreover, it is worth underlining that the concentration of EOs strongly influenced not only the growth of fungi, but also the mycotoxin biosynthesis. As the results show, the higher concentrations of EOs completely inhibited the growth of fungi, however, the influence of the lower concentration (0.1%) on the fungal growth was also evinced by ERG reduction. The volatilization method showed only the effects of higher doses of EOs. Some authors also pay attention to the dependence of the degree of fungal growth inhibition and mycotoxin level decrease on the EO concentration. Wan et al. [48] used different concentrations of five encapsulated EOs (concentration in rice culture: 1250 μg/g of bulk EOs, 1250 μg/g of EO nanoemulsions and 125 μg/g of 10-fold diluted EO nanoemulsions) to study their efficacy in the inhibition of *F. graminearum* growth and mycotoxin (DON, 3ADON, 15ADON) production. The authors stated that inhibition of mycelium growth and reduction in mycotoxin production were dose-dependent, similarly to the present work. Olosunde et al. [49] observed that with the increasing dose of *Aframomum danielli*, the degree of fumonisins reduction was higher. Application of 0.25% EO caused an 8% reduction in FB1, while the dose of 10% resulted in 76% reduction. Sumalan [50] also reported that a reduction in fumonisins was dose-dependent. In this study, it should be noted that the degree of reduction in toxin amount was not always consistent with the degree of ERG reduction. It is especially important to take into account the design of the appropriate amount of EO to apply. While the reduction in ZEA was consistent with the inhibition of ERG biosynthesis, the reduction in DON was not consistent with this parameter. In the research of da Silva Bomfim et al. [51] concerning the antifungal and antimycotoxigenic properties of rosemary EO, the production of aflatoxins was inhibited at concentrations lower than the concentration demonstrating fungistatic activity. The growth of *Aspergillus flavus* was inhibited with both MIC and MFC at 500 μg/mL, while the production of aflatoxins B1 and B2 was reduced after treatment with 250 μg/mL of rosemary EO. Similar results were obtained by Rasooli & Owlia [52] with the essential oils from *Thymus eriocalyx* and *Thymus X-porlock*. Some other authors also reported antimycotoxigenic properties at lower concentrations of EO than the concentration showing fungicidal activity [53,54].

Microscopic analysis revealed that contact of fungal cells with EOs caused morphological changes, the intensity of which depended on the oil concentration. The deformations of individual hyphae were already visible in a concentration of 0 1% of EO, while concentrations of 1% and above caused a significant decrease in viability, visible in fluorescent staining and cell deformations. Fluorescence-based microscopic methods have been used for viability testing of different bacteria, yeasts, and filamentous fungi [55–58]. The combination of FDA or cFDA and PI stains is often used to observe the effect of some antimicrobial agents on the microbial cells. For example, Li et al. [59] described the antifungal effect of a novel compound, CF66I, produced by *Burkholeria cepacia* against *Fusarium solani*, using dual staining with propidium iodide (PI) and fluorescein diacetate (FDA) to observe morphological changes. Incubation with high doses of CF66I revealed collapsed hyphae, indicating cell death and membrane permeation. In the study of Kim et a. [60], who used only PI staining, fungal hyphae treated with PGP-C (lipopeptide isolated from *Paenibacillus elgii* JCK1400 culture) showed strong red color, which indicates damage of the cell membrane. PI is a DNA-intercalating fluorescent dye that can penetrate only dead cells, in response to membrane damage.

A greater knowledge and understanding the mode of action of EOs is necessary to plan the direction and form of their application, as well as to increase the scale of pro-duction. The presented characteristics of EOs, including their antifungal and antimycotoxigenic properties, clearly indicated that their activity is dependent on the concentration used. As some authors underline, application method could also affect the effectiveness of EOs. Suhr and Nielsen [61] determined the antifungal activity of some EOs, and have stated that larger compounds, such as thymol and eugenol, demonstrated the best effects against rye bread spoilage after application directly to the medium, whereas smaller compounds, such as allyl isothiocyanate and citral, were most efficient when added as volatiles. The future application should also be considered when choosing a research method and experiment conditions, such as medium or environmental factors. Some authors indicated the lack of congruence between agar media assays and real food studies [62,63] (Arras and Usai 2001; Ultee and Smid 2001). Therefore, in the presented work, the experiments that were conducted on wheat and maize grain treated with EOs simulated the potential seed treatment. Moreover, results obtained by volatilization method can be taken into account when considering the use of fumigation.

5. Conclusions

In conclusion, the data described in this study demonstrate that EOs of cinnamon bark, palmarosa, and verbena were efficient at inhibiting *F. graminearum* growth as decreased mycelium growth, and alterations in morphological structures and ergosterol quantification were observed. The current study demonstrated that all three tested EOs effectively suppressed the development of *F. graminearum* in concentrations of 5 and 10%. The in vitro fungistatic effect was dependent on the tested EOs as well as on its concentration. Among them, cinnamon and verbena EOs displayed the strongest in vitro antifungal activity against the tested fungi. Furthermore, the tested natural plant compounds proved to be effective in mycotoxin production-inhibition, and may be an alternative to synthetic chemical agents. The high importance of the tested EOs' antimycotoxigenic properties is related to the damage that these *Fusarium* species mycotoxins cause to human and animal health. Recently developed fungicides often demonstrate systemic or curative properties aimed at effective disease control, but have limited applications at low rates. Furthermore, the emerging problem of fungicide-resistant microorganisms urges researchers to develop new biocontrol agents or fungicides possessing novel action mechanisms. Therefore, cinnamon bark, palmarosa leaves, and verbena leaf and flower EOs were proved to be potential biological agents for the control of *F. graminearum* diseases in crops and ZEA and DON mycotoxin production. Future in vitro tests are mandatory, including tests on different *Fusarium* species and mycotoxins, as well as concerning EO composition, in order to evaluate the broad spectrum of fungistatic activity as well as the inhibition of varied mycotoxin production. Additional advanced greenhouses and field in vivo research are needed in order to determine the potential usefulness of palmarosa, verbena, and cinnamon bark EOs in fungal control programs.

Supplementary Materials: The following supporting information can be downloaded at: https://www.mdpi.com/article/10.3390/app12010473/s1. Figure S1. Morphology of *F. graminearum* hyphae under light microscope (magnification × 400) treated with verbena EO: (a) Control; (b) EO concentration 0.1%; (c) EO concentration 1.0%; (d) EO concentration 5.0%; (e) EO concentration 10.0%; Figure S2. Morphology of *F. graminearum* hyphae under optical microscope (magnification × 400): untreated: (a) control, stained with cFDA; (b) control, stained with PI; treated with palmarosa EO: stained with cFDA, EO concentration: (c) 0.1%; (e) 1.0%; (g) 5.0%; (i) 10.0%; stained with PI, EO concentration: (d) 0.1%; (f) 1.0%; (h) 5.0%; (j) 10.0%; Figure S3. Morphology of *F. graminearum* hyphae under optical microscope (magnification × 400): untreated: (a) control, stained with cFDA; (b) control, stained with PI; treated with verbena EO: stained with cFDA, EO concentration: (c) 0.1%; (e) 1.0%; (g) 5.0%; (i) 10.0%; stained with PI, EO concentration: (d) 0.1%; (f) 1.0%; (h) 5.0%; (j) 10.0%.

Author Contributions: Conceptualization, D.G., K.M. and K.J.; methodology, D.G, K.M., K.J. and A.W.; validation, K.J. and A.W.; formal analysis, K.M., D.G., R.G. and R.K.; investigation, D.G., K.M., K.J., P.A.U., R.G., A.W. and R.K.; resources, D.G., K.M., K.J. and R.G.; data curation, K.J. and A.W.; writing—original draft preparation, D.G., K.M., K.J. and A.W.; writing—review and editing, D.G., K.M. and K.J.; visualization, D.G., K.J. and K.M.; supervision, D.G.; project administration, K.M. All authors have read and agreed to the published version of the manuscript.

Funding: This research was funded by Polish National Science Centre, grant number 2018/31/B/NZ9/03485.

Institutional Review Board Statement: Not applicable.

Informed Consent Statement: Not applicable.

Data Availability Statement: Not applicable.

Conflicts of Interest: The authors declare no conflict of interest.

References

1. Shahbandeh, M. Worldwide Production of Grain in 2020/21, by Type. *Statista* **2021**. Available online: https://www.statista.com/statistics/263977/world-grain-production-by-type/ (accessed on 24 September 2021).
2. Ranum, P.; Peña-Rosas, J.P.; Garcia-Casal, M.N. Global maize production, utilization, and consumption. *Ann. N. Y. Acad. Sci.* **2014**, *1312*, 105–112. [CrossRef] [PubMed]
3. Deepa, N.; Nagaraja, H.; Sreenivasa, M.Y. Prevalence of fumonisin producing *Fusarium verticillioides* associated with cereals grown in Karnataka (India). *Food Sci. Hum. Wellness* **2016**, *5*, 156–162. [CrossRef]
4. Ferreira, F.M.D.; Hirooka, E.Y.; Ferreira, F.D.; Silva, M.V.; Mossini, S.A.G.; Machinski, M., Jr. Effect of *Zingiber officinale* Roscoe essential oil in fungus control and deoxynivalenol production of *Fusarium graminearum* Schwabe in vitro. *Food Addit. Contam. Part A* **2018**, *35*, 2168–2174. [CrossRef]
5. Lofgren, L.; Riddle, J.; Dong, Y.; Kuhnem, P.R.; Cummings, J.A.; Del Ponte, E.M.; Bergstrom, G.C.; Kistler, H.C. A high proportion of NX-2 genotype strains are found among *Fusarium graminearum* isolates from northeastern New York State. *Eur. J. Plant Pathol.* **2018**, *150*, 791–796. [CrossRef]
6. Karlovsky, P. Biological detoxification of the mycotoxin deoxynivalenol and its use in genetically engineered crops and feed additives. *Appl. Microbiol. Biotechnol.* **2011**, *91*, 491–504. [CrossRef] [PubMed]
7. Dweba, C.C.; Figlan, S.; Shimelis, H.A.; Motaung, T.E.; Sydenham, S.; Mwadzingeni, L.; Tsilo, T.J. Fusarium head blight of wheat: Pathogenesis and control strategies. *Crop Prot.* **2017**, *91*, 114–122. [CrossRef]
8. Palacios, S.A.; Erazo, J.G.; Ciasca, B.; Lattanzio, V.M.T.; Reynoso, M.M.; Farnochi, M.C.; Torres, A.M. Occurrence of deoxynivalenol and deoxynivalenol-3-glucoside in durum wheat from Argentina. *Food Chem.* **2017**, *230*, 728–734. [CrossRef]
9. Escrivá, L.; Font, G.; Manyes, L. In vivo toxicity studies of fusarium mycotoxins in the last decade: A review. *Food Chem. Toxicol.* **2015**, *78*, 185–206. [CrossRef]

10. Kelly, A.C.; Clear, R.M.; O'Donnell, K.; McCormick, S.; Turkington, T.K.; Tekauz, A.; Gilbert, J.; Kistler, H.C.; Busman, M.; Ward, T.J. Diversity of Fusarium head blight populations and trichothecene toxin types reveals regional differences in pathogen composition and temporal dynamics. *Fungal Genet. Biol.* **2015**, *82*, 22–31. [CrossRef]
11. Martínez-Romero, D.; Serrano, M.; Bailén, G.; Guillén, F.; Zapata, P.J.; Valverde, J.M.; Castillo, S.; Fuentes, M.; Valero, D. The use of a natural fungicide as an alternative to preharvest synthetic fungicide treatments to control lettuce deterioration during postharvest storage. *Postharvest Biol. Technol.* **2008**, *47*, 54–60. [CrossRef]
12. Boukaew, S.; Prasertsan, P.; Sattayasamitsathit, S. Evaluation of antifungal activity of essential oils against aflatoxigenic *Aspergillus flavus* and their allelopathic activity from fumigation to protect maize seeds during storage. *Ind. Crops Prod.* **2017**, *97*, 558–566. [CrossRef]
13. Sumalan, R.M.; Alexa, E.; Popescu, I.; Negrea, M.; Radulov, I.; Obistioiu, D.; Cocan, I. Exploring Ecological Alternatives for Crop Protection Using *Coriandrum sativum* Essential Oil. *Molecules* **2019**, *24*, 2040. [CrossRef] [PubMed]
14. Tongnuanchan, P.; Benjakul, S. Essential Oils: Extraction, Bioactivities, and Their Uses for Food Preservation. *J. Food Sci.* **2014**, *79*, R1231–R1249. [CrossRef] [PubMed]
15. Fornari, T.; Vicente, G.; Vázquez, E.; García-Risco, M.R.; Reglero, G. Isolation of essential oil from different plants and herbs by supercritical fluid extraction. *J. Chromatogr. A* **2012**, *1250*, 34–48. [CrossRef]
16. Ayvaz, A.; Sagdic, O.; Karaborklu, S.; Ozturk, I. Insecticidal Activity of the Essential Oils from Different Plants Against Three Stored-Product Insects. *J. Insect Sci.* **2010**, *10*, 1–13. [CrossRef] [PubMed]
17. Bertuzzi, G. Antioxidative Action of *Citrus limonum* Essential Oil on Skin. *Eur. J. Med. Plants* **2013**, *3*, 1–9. [CrossRef]
18. Properzi, A.; Angelini, P.; Venanzoni, R. Some Biological Activities of Essential Oils. *Med. Aromat. Plants* **2013**, *2*, 1000136. [CrossRef]
19. Prakash, B.; Kedia, A.; Mishra, P.K.; Dubey, N.K. Plant essential oils as food preservatives to control moulds, mycotoxin contamination and oxidative deterioration of agri-food commodities—Potentials and challenges. *Food Control* **2015**, *47*, 381–391. [CrossRef]
20. Gakuubi, M.M.; Maina, A.W.; Wagacha, J.M. Antifungal Activity of Essential Oil of *Eucalyptus camaldulensis* Dehnh. against Selected *Fusarium* spp. *Int. J. Microbiol.* **2017**, *2017*. [CrossRef]
21. Sharma, A.; Rajendran, S.; Srivastava, A.; Sharma, S.; Kundu, B. Antifungal activities of selected essential oils against *Fusarium oxysporum* f. sp. lycopersici 1322, with emphasis on *Syzygium aromaticum* essential oil. *J. Biosci. Bioeng.* **2017**, *123*, 308–313. [CrossRef]
22. Roselló, J.; Sempere, F.; Sanz-Berzosa, I.; Chiralt, A.; Santamarina, M.P. Antifungal activity and potential use of essential oils against *Fusarium culmorum* and *Fusarium verticillioides*. *J. Essent. Oil-Bear. Plants* **2015**, *18*, 359–367. [CrossRef]
23. Kalagatur, N.K.; Nirmal Ghosh, O.S.; Sundararaj, N.; Mudili, V. Antifungal Activity of Chitosan Nanoparticles Encapsulated With *Cymbopogon martinii* Essential Oil on Plant Pathogenic Fungi *Fusarium graminearum*. *Front. Pharmacol.* **2018**, *9*, 610. [CrossRef] [PubMed]
24. Soylu, E.M.; Soylu, S.; Kurt, S. Antimicrobial activities of the essential oils of various plants against tomato late blight disease agent *Phytophthora infestans*. *Mycopathologia* **2006**, *161*, 119–128. [CrossRef] [PubMed]
25. Perumal, A.B.; Sellamuthu, P.S.; Nambiar, R.B.; Sadiku, E.R. Antifungal activity of five different essential oils in vapour phase for the control of *Colletotrichum gloeosporioides* and *Lasiodiplodia theobromae* in vitro and on mango. *Int. J. Food Sci. Technol.* **2016**, *51*, 411–418. [CrossRef]
26. Brul, S.; Nussbaum, J.; Dielbandhoesing, S.K. Fluorescent probes for wall porosity and membrane integrity in filamentous fungi. *J. Microbiol. Methods* **1997**, *28*, 169–178. [CrossRef]
27. Chitarra, G.S.; Breeuwer, P.; Nout, M.J.R.; Van Aelst, A.C.; Rombouts, F.M.; Abee, T. An antifungal compound produced by *Bacillus subtilis* YM 10–20 inhibits germination of *Penicillium roqueforti* conidiospores. *J. Appl. Microbiol.* **2003**, *94*, 159–166. [CrossRef] [PubMed]
28. Perczak, A.; Gwiazdowska, D.; Marchwińska, K.; Juś, K.; Gwiazdowski, R.; Waśkiewicz, A. Antifungal activity of selected essential oils against *Fusarium culmorum* and *F. graminearum* and their secondary metabolites in wheat seeds. *Arch. Microbiol.* **2019**, *201*, 1085–1097. [CrossRef]
29. Perczak, A.; Gwiazdowska, D.; Gwiazdowski, R.; Juś, K.; Marchwińska, K.; Waśkiewicz, A. The Inhibitory Potential of Selected Essential Oils on *Fusarium* spp. Growth and Mycotoxins Biosynthesis in Maize Seeds. *Pathogens* **2020**, *9*, 23. [CrossRef]
30. Breeuwer, P.; Drocourt, J.L.; Bunschoten, N.; Zwietering, M.H.; Rombouts, F.M.; Abee, T. Characterization of uptake and hydrolysis of fluorescein diacetate and carboxyfluorescein diacetate by intracellular esterases in *Saccharomyces cerevisiae*, which result in accumulation of fluorescent product. *Appl. Environ. Microbiol.* **1995**, *61*, 1614–1619. [CrossRef]
31. Oparka, K.J. Uptake and compartmentation of fluorescent probes by plant cells. *J. Exp. Bot.* **1991**, *42*, 565–579. [CrossRef]
32. Eskola, M.; Kos, G.; Elliott, C.T.; Hajšlová, J.; Mayar, S.; Krska, R. Worldwide contamination of food-crops with mycotoxins: Validity of the widely cited 'FAO estimate' of 25%. *Crit. Rev. Food Sci. Nutr.* **2020**, *60*, 2773–2789. [CrossRef]
33. Chilaka, C.; De Boevre, M.; Atanda, O.; De Saeger, S. Occurrence of Fusarium Mycotoxins in Cereal Crops and Processed Products (Ogi) from Nigeria. *Toxins* **2016**, *8*, 342. [CrossRef]
34. Singh, R.; Singh, G.S. Traditional agriculture: A climate-smart approach for sustainable food production. *Energy Ecol. Environ.* **2017**, *2*, 296–316. [CrossRef]
35. Żmija, D. Zrównoważony rozwój rolnictwa i obszarów wiejskich w Polsce. *Studia Ekon.* **2014**, *166*, 149–158.

36. Garcia, S.N.; Osburn, B.I.; Jay-Russell, M.T. One Health for Food Safety, Food Security, and Sustainable Food Production. *Front. Sustain. Food Syst.* **2020**, *4*, 1. [CrossRef]
37. Patriarca, A.; Fernández Pinto, V. Prevalence of mycotoxins in foods and decontamination. *Curr. Opin. Food Sci.* **2017**, *14*, 50–60. [CrossRef]
38. Čolović, R.; Puvača, N.; Cheli, F.; Avantaggiato, G.; Greco, D.; Đuragić, O.; Kos, J.; Pinotti, L. Decontamination of Mycotoxin-Contaminated Feedstuffs and Compound Feed. *Toxins* **2019**, *11*, 617. [CrossRef]
39. Bubici, G.; Kaushal, M.; Prigigallo, M.I.; Cabanás, C.G.L.; Mercado-Blanco, J. Biological control agents against Fusarium wilt of banana. *Front. Microbiol.* **2019**, *10*, 616. [CrossRef]
40. Singh, P.; Pandey, A.K. Prospective of essential oils of the genus mentha as biopesticides: A review. *Front. Plant Sci.* **2018**, *9*, 1295. [CrossRef] [PubMed]
41. Massoud, M.A.; Saad, A.S.A.; Soliman, E.A.; El-Moghazy, A.Y. Antifungal activity of some essential oils applied as fumigants against two stored grains fungi. *J. Adv. Agric. Res.* **2012**, *17*, 296–306.
42. Kowalska, J.; Tyburski, J.; Krzymińska, J.; Jakubowska, M. Cinnamon powder: An in vitro and in vivo evaluation of antifungal and plant growth promoting activity. *Eur. J. Plant Pathol.* **2020**, *156*, 237–243. [CrossRef]
43. Wu, D.; Lu, J.; Zhong, S.; Schwarz, P.; Chen, B.; Rao, J. Influence of nonionic and ionic surfactants on the antifungal and mycotoxin inhibitory efficacy of cinnamon oil nanoemulsions. *Food Funct.* **2019**, *10*, 2817–2827. [CrossRef] [PubMed]
44. Rafiee, H.; Mehrafarin, A.; Omidi, H.; Badi, H.N.; Khalighi-Sigaroodi, F. Evaluation of Phytochemical and Biochemical Patterns of Lemon Verbena (*Lippia citriodora* H.B.K.) at Different Temperatures. *Res. J. Pharmacogn.* **2019**, *6*, 9–24. [CrossRef]
45. Hosseini, M.; Jamshidi, A.; Raeisi, M.; Azizzadeh, M. The Antibacterial and Antioxidant Effects of Clove (*Syzygium aromaticum*) and Lemon Verbena (*Aloysia citriodora*) Essential Oils. *J. Hum. Environ. Health Promot.* **2019**, *5*, 86–93. [CrossRef]
46. Krzyśko-Łupicka, T.; Sokół, S.; Piekarska-Stachowiak, A. Evaluation of Fungistatic Activity of Eight Selected Essential Oils on Four Heterogeneous Fusarium Isolates Obtained from Cereal Grains in Southern Poland. *Molecules* **2020**, *25*, 292. [CrossRef]
47. da Silva Franca, K.R.; dos Santos, X.A.L.; de Alves, F.M.F.; Lima, T.S.; de Araújo, I.G.; da Nóbrega, L.P. Control of Fusarium verticillioides using Palmarosa Essential Oil (*Cymbopogon martinii*) Maracajá View project Ecology of waterbirds in semi-arid lands View project. *Artic. Int. J. Curr. Microbiol. Appl. Sci.* **2019**, *8*, 484–494. [CrossRef]
48. Wan, J.; Zhong, S.; Schwarz, P.; Chen, B.; Rao, J. Physical properties, antifungal and mycotoxin inhibitory activities of five essential oil nanoemulsions: Impact of oil compositions and processing parameters. *Food Chem.* **2019**, *291*, 199–206. [CrossRef] [PubMed]
49. Olosunde, O.O.; Adegoke, G.O.; Abiodun, O.A. Composition of sorghum-millet flour, *Aframomum danielli* essential oil and their effect on mycotoxins in kunu zaki. *Afr. J. Food Sci.* **2015**, *9*, 411–416. [CrossRef]
50. Sumalan, R.N.; Ersilia, A.; Mariana, A.P. Assessment of inhibitory potential of essential oils on natural mycoflora and fusarium mycotoxins production in wheat. *Chem. Cent. J.* **2013**, *7*, 32. [CrossRef]
51. da Silva Bomfim, N.; Kohiyama, C.Y.; Nakasugi, L.P.; Nerilo, S.B.; Mossini, S.A.G.; Romoli, J.C.Z.; Mikcha, J.M.G.; de Abreu Filho, B.A.; Machinski, M., Jr. Antifungal and antiaflatoxigenic activity of rosemary essential oil (*Rosmarinus officinalis* L.) against *Aspergillus flavus*. *Food Addit. Contam. Part A* **2020**, *37*, 153–161. [CrossRef]
52. Rasooli, I.; Owlia, P. Chemoprevention by thyme oils of *Aspergillus parasiticus* growth and aflatoxin production. *Phytochemistry* **2005**, *66*, 2851–2856. [CrossRef]
53. Tian, J.; Huang, B.; Luo, X.; Zeng, H.; Ban, X.; He, J.; Wang, Y. The control of *Aspergillus flavus* with *Cinnamomun jensenianum* hand.—Mazz essential oil and its potential use as a food preservative. *Food Chem.* **2012**, *130*, 520–527. [CrossRef]
54. Ferreira, F.D.; Kemmelmeier, C.; Arroteia, C.C.; Costa, C.L.; Mallmann, C.A.; Janeiro, V.; Ferreira, F.M.D.; Mossini, S.A.G.; Silva, E.L.; Machinski, M., Jr. Inhibitory effect of the essential oil of *Curcuma longa* L. and curcumin on aflatoxin production by *Aspergillus flavus* link. *Food Chem.* **2013**, *136*, 789–793. [CrossRef]
55. Kessel, G.J.T.; de Hass, B.H.; Lombaers-van der Plas, C.H.; Meijer, E.M.J.; Dewey, F.M.; Gou-driaan, J.; ven der Werf, W.; Kohl, J. Quantification of mycelium of *Botrytis* spp. and the antagonist *Ulocladium atrum* in necrotic leaf tissue of Cyclamen and Lily by fluorescence microscopy and image analysis. *Phytopathology* **1999**, *89*, 868–876. [CrossRef]
56. Foglieni, C.; Meoni, C.; Davalli, A.M. Fluorescent dyes for cell viability: An application on prefixed condition. *Histochem. Cell Biol.* **2001**, *115*, 223–229. [CrossRef]
57. Khan, M.M.T.; Pyle, B.H.; Camper, A.K. Specific and rapid enumeration of viable but nonculturable and viable-culturable gram-negative bacteria by using flow cytometry. *Appl. Environ. Microbiol.* **2010**, *76*, 5088–5096. [CrossRef]
58. Fernandez-Miranda, E.; Majada, J.; Casares, A. Efficacy of propidium iodide and FUN-1 stains for assessing viability in basidiospores of *Rhizopogon roseolus*. *Mycologia* **2017**, *109*, 350–358. [CrossRef]
59. Li, X.; Quan, C.S.; Yu, H.Y.; Wang, J.H.; Fan, S.D. Assessment of an-tifungal effects of a novel compound from *Burkholderia cepacia* against *Fusarium solani* by fluorescent staining. *World J. Microbiol. Biotechnol.* **2009**, *25*, 151–154. [CrossRef]
60. Kim, J.; Le, K.D.; Yu, N.H.; Kim, J.I.; Kim, J.C.; Lee, C.W. Structure and antifungal activity of pelgipeptins from *Paenibacillus elgii* against phytopathogenic fungi. *Pestic. Biochem. Physiol.* **2020**, *163*, 154–163. [CrossRef] [PubMed]
61. Suhr, K.I.; Nielsen, P.V. Antifungal activity of essential oils evaluated by two different application techniques against rye bread spoilage fungi. *J. Appl. Microbiol.* **2003**, *94*, 665–674. [CrossRef] [PubMed]

62. Arras, G.; Usai, M. Fungitoxic activity of 12 essential oils against four postharvest citrus pathogens: Chemical analysis of *Thymus capitatus* oil and its effect in subatmospheric pressure conditions. *J. Food Prot.* **2001**, *64*, 1025–1029. [CrossRef] [PubMed]
63. Ultee, A.; Smid, E.J. Influence of Carvacrol on growth and toxin pro-duction by *Bacillus cereus*. *Int. J Food Microbiol.* **2001**, *64*, 373–378. [CrossRef]

Article

Antioxidant, Antimicrobial and Antibiofilm Properties of *Glechoma hederacea* Extracts Obtained by Supercritical Fluid Extraction, Using Different Extraction Conditions

Daniela Gwiazdowska [1,*], Pascaline Aimee Uwineza [2], Szymon Frąk [1], Krzysztof Juś [1], Katarzyna Marchwińska [1], Romuald Gwiazdowski [3] and Agnieszka Waśkiewicz [2]

1 Department of Natural Science and Quality Assurance, Institute of Quality Science, Poznań University of Economics and Business, Niepodległości 10, 61-875 Poznań, Poland; szymon.frak@phd.ue.poznan.pl (S.F.); krzysztof.jus@ue.poznan.pl (K.J.); katarzyna.marchwinska@ue.poznan.pl (K.M.)
2 Department of Chemistry, Poznań University of Life Sciences, Wojska Polskiego 75, 60-625 Poznań, Poland; pascaline.uwineza@up.poznan.pl (P.A.U.); agnieszka.waskiewicz@up.poznan.pl (A.W.)
3 Research Centre for Registration of Agrochemicals, Institute of Plant Protection-National Research Institute, Władysława Węgorka 20, 60-318 Poznań, Poland; r.gwiazdowski@iorpib.poznan.pl
* Correspondence: daniela.gwiazdowska@ue.poznan.pl; Tel.: +48-61-856-95-36

Featured Application: Ensuring safe food and care for the health of consumers and animal welfare are among the objectives of the EU's policy and food production sectors. Moreover, there is a great need for the introduction of new, environmentally friendly technologies, including the extraction of antimicrobial substances. Supercritical fluid extraction is becoming an increasingly popular method for the recovery of bioactive compounds, representing a non-toxic, cheap, and generally recognized as safe (GRAS) technique, compared to conventional extraction methods, which often require higher temperatures and large amounts of organic solvents. The research presented here is the first to describe the biological activity of *Glechoma hederacea* extracts obtained by means of supercritical fluid extraction. Therefore, it provides new information and broadens the existing knowledge in the study of the properties of SC-CO2 plant extracts and their potential application.

Abstract: *Glechoma hederacea* var. *longituba* is a herbaceous plant from the *Lamiaceae* family, used in herbal medicine. In this work, we aimed to assess the total phenolic content, antioxidant, antimicrobial and antibiofilm activity of extracts obtained from *G. hederacea* via supercritical dioxide extraction with methanol as a co-solvent under different extraction conditions. The results showed that the activity of the obtained SC-CO_2 extracts is strongly dependent on the extraction temperature. Significantly higher total polyphenol content, as well as antioxidant and antimicrobial activity towards bacteria and yeasts, was observed in the extract obtained at 40 °C, compared to extracts obtained at 50 °C and 60 °C; however, antifungal activity against filamentous fungi was not dependent on the extraction conditions. Antimicrobial activity also depended on the microorganism type. Higher sensitivity was exhibited by Gram-positive bacteria than by Gram-negative bacteria, with *S. aureus* and *P. aeruginosa* being the most sensitive species among each group. The most susceptible fungi were *Candida albicans* and *Sclerotinia sclerotiorum*. The antibiofilm activity was differentiated and depended on the extraction conditions, the microorganism and the method of biofilm treatment. All tested extracts inhibited biofilm formation, with the extract obtained at 40 °C showing the highest value, whereas only extract obtained at 60 °C efficiently removed mature biofilm.

Keywords: antioxidant activity; antimicrobial properties; biofilm; *Glechoma hederacea*; plant extracts; supercritical fluid extraction

Citation: Gwiazdowska, D.; Uwineza, P.A.; Frąk, S.; Juś, K.; Marchwińska, K.; Gwiazdowski, R.; Waśkiewicz, A. Antioxidant, Antimicrobial and Antibiofilm Properties of *Glechoma hederacea* Extracts Obtained by Supercritical Fluid Extraction, Using Different Extraction Conditions. *Appl. Sci.* **2022**, *12*, 3572. https://doi.org/10.3390/app12073572

Academic Editor: Antonio Valero

Received: 8 February 2022
Accepted: 29 March 2022
Published: 31 March 2022

1. Introduction

Food safety is concerned with protecting the food supply chain against the introduction, development or survival of hazardous microorganisms and their metabolites, as well as chemical agents [1]. Therefore, maintaining high quality and safety of food is a public health priority. According to WHO reports [2] 18 million disability-adjusted life years (DALYs) have been lost due to foodborne pathogens worldwide, with nontyphoidal *Salmonella enterica* and enteropathogenic *Escherichia coli* being the dominant microorganisms. Research shows that non-optimal food hygiene practices may contribute to microbial contamination of food [3] and thus to bacterial food poisoning, which is one of the most common causes of illness and death all over the world [4]. The majority of food poisoning cases are linked to bacterial contamination, particularly *Salmonella* species, *Bacillus cereus*, *Staphylococcus aureus*, *Escherichia coli* species, *Clostridium* species, *Campylobacter jejuni* and *Listeria monocytogenes* [5–7]. Fungi also pose a serious risk to food safety at every stage of the food chain. Filamentous fungi such as *Fusarium*, *Alternaria*, *Botrytis* and *Sclerotinia* are responsible for plant diseases and can be dangerous to humans and animals among others due to mycotoxins production. These metabolites decrease nutritional properties and pose a threat to human and animal health by causing acute or chronic problems such as carcinogenicity, mutagenicity, teratogenicity and hepatotoxicity [8–11].

In general, chemical compounds are used at different stages of food production to prevent and control microbial contamination and spoilage. Agrotechnical procedures include the application of fungicides, whereas in the food industry disinfectants or preservatives are used. The disadvantage of chemicals is the possibility of the accumulation of their residues in the food and feed chain, the development of microbial resistance to the applied compounds and other side effects on human and animal health [4,12]. Therefore, efforts have been made to develop potentially effective, healthy, safer and natural substances as alternatives to the commonly used compounds. One solution was found in the use of plant extracts, which have been extensively studied for their antimicrobial properties [13]. For example, roselle (*Hibiscus sabdariffa*), rosemary (*Rosmarinus officinalis*), clove (*Syzygium aromaticum*), thyme (*Thymus vulgaris*) and other herbs demonstrate both antibacterial and antifungal activity against various microorganisms [14,15].

Plant extracts obtained from aromatic, medicinal or herbal plants consist of compounds that are increasingly being used as preservatives in the food industry, in pharmaceuticals and cosmetics and as natural fungicides in agriculture. *Glechoma hederacea* var. *longituba*, commonly known as ground ivy, is a herbaceous plant from the *Lamiaceae* family, that is widely available throughout Asia, Europe and North America [16]. In some regions, *G. hederacea* is known as a weed plant that grows in shaded areas, fallow lands, dry ditches, around fences and hedges and along the edges of wet meadows [17]. According to the literature, *G. hederacea* leaves or flowering herbs have long been used as a traditional medicine in the treatment of various diseases, such as abscess, arthritis, asthma, cold, cough, diabetes, influenza, gastric disorders, headaches, hypochondria, inflammation, jaundice and scurvy [17,18].

Different chemical compositions of *G. hederacea* have been described, and a variety of active compounds, such as phenolic acids, flavonoids, terpenoids, alkaloids, steroids and fatty acids, have been identified [17,19–22]. Phenolic compounds, including rosmarinic acid, chlorogenic acid, caffeic acid, rutin, genistin and ferulic acid, have been confirmed as major constituents of *G. hederacea* extracts [21,23]. Moreover, norlignans, tropane alkaloids (hederacins), sesquiterpenes, sesquiterpene lactones, triterpenoids (such as ursolic and oleanolic acids), essential oils and lectins have also been identified in *Glechoma* [22,24–26].

Literature data revealed that the biological effects of this plant include anti-melanogenic, anti-inflammatory, antibacterial, antimutagenic, genoprotective, antigenotoxic and antitumor effects [17,18,27–29]. Moreover, its antioxidant properties have been reported with respect to its use in the food industry [26,30]. According to the literature data, *Glechoma* is a promising source of bioactive constituents that can be beneficial in a sustainable manner, acting as natural antioxidants and antibacterial agents, but it has not been associated with phytopathogenic fungal efficacy.

Extraction methods of bioactive compounds from aromatic or medicinal plants have a significant impact on the quality of the extracts and their chemical composition [12]. The literature usually describe *G. hederacea* extracts prepared using traditional procedures such as distillation, with different parameters and solvents applied [21,26,31,32]. However, these extraction methods have a number of limitations, including the fact that they are time-consuming, labor-intensive procedures that require a lot of solvents and, in some cases, produce low yields. Therefore, new potential extraction methods have emerged in recent years that provide some type of additional energy to enable the faster transfer of solutes from the sample to the solvent. Supercritical fluid extraction (SFE) is one of the alternative methods to conventional systems that has gained acceptance in the extraction of bioactive compounds from a variety of materials [33]. It is considered a clean method due to the exceptional purity of the extracts obtained. Carbon dioxide (CO_2) is usually the preferred solvent in supercritical fluid extraction. It offers several advantages over other solvents, including its low cost, nonflammability, chemical inertness and lack of toxicity [33,34]. However, polar chemicals such as phenolics cannot be extracted directly using supercritical-CO_2 (SC-CO_2) due to the non-polarity of CO_2. Therefore, methanol, water or ethanol are added as co-solvents or modifiers to improve the solvation power, affinity for weakly soluble solutes (alkaloids, phenolics and glycosidic chemicals), solubility and extraction yield, depending on the operating pressure and temperature [35]. When compared to conventional separation techniques, SFE with carbon dioxide (SC-CO_2) has several advantages, including physicochemical properties that are halfway between a liquid and a gas, with low viscosity, high density and diffusivity; the fact that thermally sensitive compounds can be separated at low temperatures; the solvent can be easily removed from the extracts through pressure reduction or temperature elevation; and that some studies have indicated that SC-CO_2 extracts have the highest antioxidant and antifungal activities [34–37].

In the relation to the data mentioned above, the aim of the present work was to investigate the total phenolic content and antioxidant, antimicrobial and antibiofilm activity of *G. hederacea* extracts obtained via supercritical dioxide extraction with methanol as the co-solvent under differential extraction conditions.

2. Materials and Methods

2.1. Material and Sample Extraction

Dried, ground *G. hederacea* var. *longituba* herbs were purchased from a Polish manufacturer of high-quality natural herbal products, FLOS Elżbieta and Jan Głąb Spółka Jawna, Poland. The experiment was performed using a supercritical fluid extraction system, specifically, SC-CO_2, following the procedure described by Uwineza et al. (2021) [38]. The grounded and dried *G. hederacea* herbs of 5 g were placed in an extraction vessel of 25 mL and kept in an oven set at different temperatures (40, 50, and 60 °C) and constant pressure (250 bar). The CO_2 flow rate was set to 4 mL/min and 1 mL/min of pure methanol (99.5% purity) was used as a co-solvent. The extraction process was started automatically after the system reached the established conditions and was carried out for 180 min in each experimental run which was composed of 1st dynamic time—45 min, static time—15 min and 2nd dynamic time—120 min. *G. hederacea* extracts were collected in flasks placed in a fraction collection module, and stored at −20 °C for further analysis.

2.2. Chemicals

Methanol for the HPLC-super gradient was purchased from POCh (Gliwice, Poland) and Folin–Ciocalteu's reagent and hydrochloric acid of 35–38% purity were purchased from Chempur (Piekary Śląskie, Poland). 3,4,5-trihydroxybenzoic acid (gallic acid), 2,2-dyphenyl-1-picrylhydrazyl (DPPH), iron (III) chloride, sodium acetate, sodium carbonate anhydrous, potassium acetate, acetic acid glacial, phosphate buffered saline tablet, potassium persulfate,

2,4,6-tris(2-pyridyl)-s-triazine (TPTZ), 2,2′-azino-bis(3-ethylbenzothiazoline-6-sulfonic acid (ABTS) and (±)-6-hydroxy-2,5,7,8-tetramethylchromane-2-carboxylic acid (Trolox) were purchased from Sigma-Aldrich (Poland). Carbon dioxide (CO_2, SFE grade), was purchased from Air Products Sp, Poland. Microbiological media were purchased from BioMaxima (Poland) and A&A Biotechnology (Poland). All chemicals were of analytical grade.

2.3. Total Phenolic Content of *G. hederacea* Extracts

The total phenolic content (TPC) of *G. hederacea* extracts was measured using the Folin–Ciocalteu assay [39]. In a test tube, 1.60 mL of distilled water was mixed with 20 µL of the sample extract, blank or standard to be analyzed. After that, 100 µL of the Folin–Ciocalteu reagent was added and vortexed. After 3 min, 300 µL of 75 g/L Na_2CO_3 was added and stirred vigorously. After incubating the solution at room temperature for 45 min in the absence of light, absorbance values at 760 nm were acquired using a Varian Cary 300bio UV-VIS spectrophotometer (Agilent, Santa Clara, CA, USA). Gallic acid was used as a standard and a calibration curve was plotted in the range 50–500 mg/mL [40] and the blank sample was distilled water. All measurements were performed at least in triplicate and the total phenolic content estimation was calculated using the following formula according to Mabrouki et al. (2018) [41]:

$$\mathrm{TPC} = \mathrm{c} \times \mathrm{v}/\mathrm{m} \tag{1}$$

where c is the concentration of gallic acid established from the calibration curve (mg/mL), v is the volume of extract solution (mL) and m is the weight of the sample extract (g). The results were expressed as mg gallic acid equivalent per gram of the extract (mg GAE/g of extract).

2.4. Evaluation of Antioxidant Activity

2.4.1. Ferric Reducing Antioxidant Power (FRAP) Assay

A modified version of the FRAP assay was used to assess the ferric reducing capacity of *G. hederacea* extracts [42]. The reduction of a pale-yellow-colored ferric complex (Fe^{3+}-tripyridyl triazine) to a blue-colored ferrous complex (Fe^{2+}-tripyridyl triazine) by electron-donating antioxidants at low pH is the basis for this approach. The working FRAP reagent was prepared daily by mixing 10 mL of 300 mM acetate buffer of 3.6 pH, with 1 mL of 10 mM TPTZ (2,4,6-tri(2-pyridyl)-s-triazine) in 40 mM hydrochloric acid and with 1 mL of 20 mM ferric chloride. The reaction mixture was incubated for 15 min at 37 °C in a water bath before use. An aliquot of 3 mL of the freshly prepared FRAP reagent was added to 100 µL of plant extracts or standard (Trolox) and incubated for 5 min at 37 °C in a water bath before analysis. Then, the absorbance of the samples was measured at 593 nm using a Varian Cary 300bio UV-VIS spectrophotometer (Agilent, Santa Clara, CA, USA). All measurements were carried out in triplicate. Trolox was used as the standard and the calibration curve was plotted in the range of 150–3000 µmol/L [42]. The FRAP value was determined as previously reported [38]. Based on the obtained FRAP value, the final antioxidant activity (AA) in each sample was expressed as a Trolox equivalent (TE) in terms of µmol Trolox equivalent TE/g of extract, according to the following equation [43]:

$$\mathrm{AA}\ (\mu\mathrm{mol\ TE/g\ extract}) = \mathrm{FRAP\ value}\ (\mu\mathrm{mol/L})/\mathrm{sample}\ (\mathrm{g/L}) \tag{2}$$

2.4.2. Free Radical Scavenging by ABTS Assay

The free radical scavenging capacity of *G. hederacea* extracts was also studied using the ABTS radical cation decolorization assay [44], which is based on the reduction of $ABTS^+$ radicals by the antioxidants of the tested plant extracts. The $ABTS^+$ radical was prepared by mixing equal amounts of two stock solutions (7 mM ABTS solution and 2.45 mM potassium persulfate solution) and allowing them to react for 16 h at room temperature in the dark. The working solution was then prepared by mixing 3.9 mL of $ABTS^+$ with 140 mL of 5 mM phosphate buffered saline (pH 7.4) to obtain an absorbance of (0.70 ± 0.02) at 734 nm using a spectrophotometer. Fresh ABTS working solution was prepared daily. The amount of

2850 μL of ABTS working solution was allowed to react with 150 μL of plant extracts in a test tube for 8 min in water bath incubation at 30 °C. The absorbance was measured at 734 nm using a Varian Cary 300bio UV-VIS spectrophotometer (Agilent, Santa Clara, CA, USA). Trolox was used as the standard and PBS as the blank. The calibration curve was plotted in the range of 0.9–10 μg/mL. The results are expressed in μg Trolox equivalents/g of extract mass (μg(TE)/g) based on the calibration curve according to the following formula:

$$\text{ABTS value } (\mu\text{g TE/g of extract}) = (c \times v)/m \quad (3)$$

where c is the Trolox concentration (μg/mL) of the corresponding standard curve of the plant extract, v is the sample volume (mL) and m is the weight of the plant extract (g) [45].

2.4.3. Free Radical Scavenging Ability by DPPH Assay

The antioxidant activity of *G. hederacea* extracts was determined based on the free radical scavenging activity of the DPPH assay (2,2-diphenyl-1-picrylhydrazyl radical), which was modified slightly from the method described by Moradi et al. (2016) [46]. The working solution was prepared by preparing a methanolic solution of DPPH (0.1 mM). An aliquot of 2850 μL of this solution was mixed with 150 μL of the sample, the standard (Trolox) under different concentrations or the blank (methanol). The reaction mixture was thoroughly mixed before incubation in the dark for 30 min at room temperature. After that, the absorbance at 517 nm was measured using a Varian Cary 300bio UV-VIS spectrophotometer (Agilent, Santa Clara, CA, USA). Low absorbance of the reaction mixture indicated high free radical scavenging activity. The experiment was repeated three times, and the findings presented are averages of the three values. The scavenging activity was estimated based on the percentage of DPPH radical scavenged according to the following equation:

$$\text{DPPH inhibition \%} = (B - S/B) \times 100 \quad (4)$$

where B is the absorbance of the blank and S is the absorbance of the sample.

2.5. Antimicrobial Activity of G. hederacea Extracts

2.5.1. Indicator Microorganisms

In the experiment, four Gram-positive bacteria—*Micrococcus luteus* ATCC 10240, *Staphylococus aureus* ATCC 33862, *Bacillus subtilis* ATCC 11774 and *Enterococcus faecalis* ATCC 19433—as well as three Gram-negative bacteria—*Escherichia coli* ATCC 8739, *Pseudomonas aeruginosa* ATCC 9027 and *Salmonella enterica* ser. Enteritidis ATCC 13076—were used. All bacterial strains were purchased from the American Type Culture Collection (ATCC) and cultivated on liquid media: trypticasein soy broth (TSB) for *M. luteus*, nutrient broth (NB) for *S. aureus*, *B. subtilis*, *E. coli*, *P. aeruginosa* and brain heart infusion (BHI) for *E. faecalis* and *S. enteritidis* under optimal temperature conditions (30 °C for *M. luteus* and 37 °C for the remaining bacteria). The fungistatic activity of the tested extracts was determined against one strain of yeast and five filamentous fungi. The yeast strain (*C. albicans* ATCC 10231) was purchased from the American Type Culture Collection and cultured on Sabouraud Dextrose Broth (SAB) at 37 °C under aerobic conditions. Two species of the genus *Fusarium* (*F. graminearum* KZF 1 and *F. culmorum* KZF 5), *Alternaria alternata* KZF 13 and *Sclerotinia sclerotiorum* KZF 23 were obtained from the collection of the Research Centre for Registration of Agrochemicals, whereas *Botrytis cinerea* BPR 187 was from the Bank of Plant Pathogens and Research on their Biodiversity, Institute of Plant Protection, National Research Institute in Poznań, Poland. The tested filamentous fungi were cultivated in Petri dishes (55 mm diameter) on a Potato Dextrose Agar (PDA) at 25 °C for 5–10 days.

2.5.2. Inoculum Preparation and Standardization

Bacteria and yeasts were cultured for 24 h on agar media (according to Section 2.5.1). The bacteria and yeast inocula were prepared in Mueller–Hinton broth (MHB) for bacteria and SAB for yeasts, with optical density adjusted to 0.5 McFarland standard. In the case of filamentous fungi, hyphae and conidia suspensions were prepared in sterile PDB by mixing harvested mycelium from mature cultures with medium to achieve a final cell concentration of 10^6 cells/mL, determined with a hemocytometer.

2.5.3. Minimal Inhibitory Concentration (MIC), Minimal Bactericidal Concentration (MBC), Minimal Fungicidal Concentration (MFC) Determination

The MIC, as well as the MBC/MFC, of the tested *G. hederacea* extracts were determined using the microdilution method according to Gwiazdowski et al. (2018) [47] and Rzemieniecki et al. (2019) [48] with some modifications. Twofold dilutions of the extracts were prepared in 96-well microtiter plates in MHB for bacteria, SAB for yeast and PDB for filamentous fungi. The final concentration of tested extracts was established in the range of 0.04–5 mg/mL; in the case of *Fusarium* species the range was 0.08–10 mg/mL. The final concentration of methanol in control samples was established in the range of 0.2–25% in proportion to its content in the samples. Next, 100 µL of the microorganism solutions were added to each well. The plates inoculated with bacteria and yeasts were covered and incubated for 24 h at 30 °C or 37 °C, depending on the microorganism. In the case of filamentous fungi, microtiter plates were sealed with parafilm (to minimize the risk of extracts evaporation) and incubated at 25 °C ± 2 °C for 5–10 days under aerobic conditions. Culture media containing *G. hederacea* extracts without microbial inoculum were used as negative controls, whereas bacterial or fungal cultures without extracts were used as positive controls. After incubation, the optical density of the bacterial and yeasts samples was determined at a 600 nm wavelength using the BioTek Epoch 2 microplate reader. The MIC value was defined as the lowest concentration of extract that exhibited at least 90% growth inhibition. The MBC/MFC value was determined via spot inoculation of 10 µL of microbial culture with the addition of an extract at a concentration equal to or higher than the MIC value (100% inhibition based on spectrophotometric measurements using BioTek Epoch 2). MIC/MFC values for filamentous fungi were determined through a visual assessment of the fungal growth on the plate. All tests were performed in triplicate.

2.6. Antibiofilm Activity of G. hederacea Extracts

2.6.1. Biofilm Formation

Biofilm formation experiments for selected bacteria: *P. aeruginosa*, *B. subtilis*, *E. coli* and *E. faecalis* were carried out according to the modified Somrani (2020) [49] method. Standardized bacteria cultures of 10^6 CFU/mL were prepared in the appropriate broth medium (TSB, BHI or NB) and in amounts of 60 µL added into each well. Samples were prepared in triplicate and incubated at 37 °C for 24 h. After incubation, the suspension was carefully removed from the plate and the wells were rinsed three times with water to remove non-adherent cells and the residual medium. The plates were air-dried for 2 h.

2.6.2. Assessment of Antibiofilm Activity of *G. hederacea* Extracts

The antibiofilm activity of *G. hederacea* extracts was examined in two ways: as a factor preventing biofilm formation and as a biofilm removal factor. In the first case, the extracts were added to the wells before microbial incubation, whereas in second case the extracts were used after biofilm formation to remove them. To determine the effect of the tested extracts on the ability to form bacterial biofilms, solutions of plant extracts were prepared. Into each well of flat-bottom 96-well microtiter plates, 60 µL of each extract and 60 µL of the bacterial suspension, prepared as described above, were added. The plates were incubated for 24 h at 37 °C. The final concentrations of the tested extracts were equal to the MIC values. Methanol was added as a negative control. After incubation, the suspension was removed from the plates and the wells were rinsed three times. The plates were air-dried for 2 h.

To determine the effect of the tested plant substances on the removal of the mature biofilm, the biofilm was first formed by adding 60 µL of water and 60 µL of the bacterial suspension into each well and incubating for 24 h at 37 °C. After washing with water and air-drying, the biofilm was washed three times with 125 µL of the extract at a corresponding MIC value concentration at room temperature (25 °C ± 1 °C). After 15 min, the tested substances were removed and the plates were washed with water and air-dried for 2 h.

2.6.3. Biofilm Staining and Quantifying

Biofilm biomass was determined using the modified crystal violet method developed by O'Toole (2011) [50]. The dried plates with formed biofilms were flooded with 125 µL of a 0.1% crystal violet solution for 15 min. The plates were then washed again with water and dried overnight. For the biofilm quantification, 125 µL of 30% acetic acid was added into each well and left at room temperature for 10–15 min. The contents of the wells were transferred to a new microtiter plate and the optical density of each well was analyzed spectrophotometrically at a 550 nm wavelength using a BioTek Epoch 2 microplate reader. As a blank, 30% acetic acid in water was used. Samples were conducted in triplicate parallel repetitions. Results were expressed as a percentage of inhibition of biofilm formation.

2.7. Statistical Analysis

The experimental data concerning polyphenol content and antioxidant activity were statistically evaluated using the Statgraphics 4.1 software package (Graphics Software System, STCC, Inc., Rockville, MD, USA). A one-way ANOVA was used to assess the significance of differences in antioxidant activity and polyphenol concentration in the tested extracts. Fisher's least significant difference (LSD) test at $\alpha = 0.05$ was used for the paired tests. The effect of the tested extracts on the formation and removal of bacterial biofilms was estimated via a one-way analysis of variance (ANOVA) using the IBM SPSS Statistics program. Homogeneity of variance was tested using Levene's test. Furthermore, for homogeneous samples Tukey's test was applied and for nonhomogeneous samples the Games–Howell test with a p-value < 0.05 was applied.

3. Results

3.1. Phenolic Content of G. hederacea Extracts Obtained at Different Extraction Variants

G. hederacea CO_2 extracts obtained with different extraction conditions were tested for their biological activity. The yield of SC-CO_2 extraction with methanol as a co-solvent was established in the presented work at 9.58%. In SC-CO_2 extraction, density is extremely important and highly dependent on temperature and pressure variation [51,52]. An increase in pressure favors an increase in density at a constant temperature, whereas an increase in temperature decreases the density at a constant pressure. In our study, we tested three temperatures (40, 50 and 60 °C) during the extraction process at constant pressure. At the lowest temperature we obtained the highest level of total phenolic content. As reported Bezerra et al. (2020) [52] the increase in the temperature causes a greater intermolecular distance, and consequently the reduction of the solubility power of CO_2, by decreasing the density.

In this study, the TPC in the obtained *G. hederacea* extracts was analyzed spectrophotometrically using the Folin–Ciocalteu method. The results showed that the extract obtained in conditions of 40 °C/250 bar had the highest TPC value (138.33 ± 5.00 mg GAE/g) compared with the samples obtained at temperatures of 50 °C and 60 °C (Table 1). The TPC values of the other two tested *G. hederacea* extracts were lower than $^1/_3$ of the value obtained at 40 °C (43.00 ± 3.04 mg GAE/g for 50 °C and 46.00 ± 9.26 mg GAE/g for 60 °C).

Table 1. Total phenolic content of *G. hederacea* extracts obtained from different extraction variants.

Extraction Conditions	TPC (mg GAE/g Extract)
40 °C	138.33 ^{a} ± 5.00
50 °C	43.00 ^{b} ± 3.04
60 °C	46.00 ^{b} ± 9.26

Values are mean ± standard deviation, n = 3, values with the same lowercase letters in the same column indicating no significant difference at the level of 5% ($p < 0.05$).

The most reported method for obtaining *G. hederacea* extracts is aqueous extraction and the total phenolic content obtained in this study cannot be directly compared to those extracts, as different parameters (extraction method, time, pressure, solvent) had an influence on the final results. However, it is worth mentioning that Varga et al. (2016) [32] reported total phenols from the aqueous *Glechoma* extract in the range of 43.9 ± 3.2–109.8 ± 5.8 mg GAE/g, whereas Hahm et al. (2021) [53] reported an average phenolic content of 14.81 ± 4.53 mg/g, and Chou et al. (2012) [18] found a total phenol content of 79.70 ± 0.193 mg GAE/g.

The literature data indicate that SC-CO_2 can be used to obtain extracts rich in phenolic compounds [54]. However, due to the nonpolar nature of CO_2, some polar cosolvents such as ethanol, methanol or ethyl acetate need to be added to increase the extraction yield of phenolic compounds via the increase in solvation power [52,55]. In the presented work, methanol was used as an example of a polar solvent due to its properties and costs. Moreover, the effect of temperature, pressure, flow rate and density of the supercritical CO_2 on the bioactive compound extraction process using different plant materials was analyzed previously [52]. According to some authors [51,56], solvent density or pressure may influence the mass yield of the extract. According to a study by García-Abarrio (2014), overall SFE yield increases with CO_2 density and co-solvent ratio [56]. Silva et al. (2021) [51] reported that at 40 °C the solvent density affected the mass yields, whereas after increasing temperature to 50 °C, the solvent density did not influence yields, but rather affected the vapor pressure of the solute. Based on the obtained results, we can conclude that a temperature increase has a negative effect on phenolic solubility during extraction due to the decrease in the solvent's density. Akowuah et al. (2009) [57] observed a decrease in the total phenolic compounds determined in an extract from *Gynura procumbens* leaves when the temperature was increased in a conventional solvent extraction system. When compared to other authors' studies, the high TPC value achieved for the obtained extracts (particularly for the fraction at 40 °C) shows that the extraction method employing the supercritical fluid extraction technique was very successful.

3.2. Antioxidant Effect of *G. hederacea* Extracts

Plants are a rich source of natural antioxidants, mainly phenolic compounds, that may delay, inhibit, or prevent oxidative processes that contribute to the deterioration of food quality or to the onset and development of degenerative diseases in the body [58]. Generally, antioxidant activity is primarily based on two chemical mechanisms: single-electron transfer and hydrogen atom transfer. However, due to the various mechanisms, reaction characteristics and variable phase localizations that are typically involved in the process, there is currently no single standardized method for determining antioxidant activity. It is worth noting that the DPPH radical is a stable free radical that is commonly used to assess antioxidants' free radical scavenging capacities [23]. In this study, the antioxidant activity of *G. hederacea* extracts obtained using SC-CO_2 with methanol as a co-solvent was analyzed spectrophotometrically using three different assays (DPPH, ABTS and FRAP).

The findings showed that the antioxidant activity of the *G. hederacea* extracts was significantly different at $p < 0.05$ between the extraction temperatures. However, any of these assays could be used for the analysis of the antioxidant activity of *Glechoma* because all tested assays confirmed 40 °C/250 bar to be the best conditions for the extraction of antioxidants in this study. The results are summarized in Table 2. Compared with other authors, Chou et al. (2012) [18] reported that the antioxidant activities of the hot water extract of *G. hederacea* (HWG) were significantly higher than those of vitamin C and Trolox in terms of superoxide anion radical-scavenging activity and Fe^{2+}-chelating ability ($p < 0.05$). Similarly, Oalđe et al. (2021) [29] investigated the antioxidant activity of methanolic, ethanolic and aqueous extracts of *Glechoma hederacea, Hyssopus officinalis, Lavandula angustifolia, Leonurus cardiaca, Marrubium vulgare* and *Sideritis scardica* (*Lamiaceae*) using several experimental models. Their findings revealed that the ethanolic extract of *G. hederacea* had the highest DPPH scavenging activity among the investigated extracts, which was comparable to that of the positive control, 2-tert-butyl-4-hydroxyanisole (BHA) [29]. Furthermore, Matkowski (2008) [59] examined the antioxidant capacity of extracts and various solvent fractions of *Glechoma hederacea* L. and *Orthosiphon stamineus* (Benth.) Kudo. The results demonstrated that the methanolic extracts of *O. stamineus* exhibited much higher activity than those of *G. hederacea* [59].

Table 2. Antioxidant activities estimated via DPPH, ABTS and FRAP assays of *G. hederacea* extracts.

Extraction Conditions	DPPH (%)	ABTS (μg TE/g)	FRAP (μmol TE/g)
40 °C	56.48 [a] ± 3.98	36.58 [a] ± 1.20	18.15 [a] ± 0.21
50 °C	25.74 [b] ± 0.43	7.60 [b] ± 0.69	13.06 [b] ± 0.04
60 °C	22.21 [c] ± 0.39	4.66 [c] ± 0.12	12.88 [b] ± 0.07

Values are mean ± standard deviation, n = 3, values with the same lowercase letters in the same column indicate no significant difference at the level of 5% ($p < 0.05$).

The correlation of total phenolic content with FRAP, ABTS and DPPH scavenging activities is shown in Figure 1a–c, respectively. The analysis revealed that the results of all three assays best correlated with the gallic acid equivalent values, estimated using the Folin–Ciocalteu method. The Pearson's correlation coefficient (r) and coefficient of determination (R^2) were the highest (r = 0.9983, R^2 = 0.9967) between total phenolic content and FRAP activity than those of total phenolic content and ABTS activity (r = 0.9938, R^2 = 0.9877), followed by total phenolic content and DPPH activity (r = 0.9926, R^2 = 0.9853). These results suggest that the total phenols in the *G. hederacea* extracts were the primary contributor to the antioxidant activities of *Glechoma* extracts obtained using the SC-CO_2 set at different temperatures.

3.3. Antibacterial Activity

The antibacterial properties of the tested *G. hederacea* extracts, expressed as MIC and MBC values, against Gram-positive and Gram-negative bacteria, are presented in Table 3. All tested extracts exhibited the inhibition of indicator bacteria; however, the results depended on both the bacterial strain and the sample extraction temperature. The strongest antagonistic effect was observed for the herb extract obtained at 40 °C/250 bar for most of the Gram-positive bacteria and one of the tested Gram-negative species. Stronger antibacterial properties were detected mainly in relation to Gram-positive bacteria, whereas the effect was weaker in relation to Gram-negative bacteria. MIC ranged from 0.3 to >5.0 mg/mL for Gram-positive bacteria, and from 1.25 to 2.5 mg/mL for Gram-negative bacteria, whereas MBC ranged from 0.6 to >5 mg/mL and from 2.5 to 5.0 mg/mL, respectively. The MIC values for all extracts against Gram-positive *M. luteus* (1.25 mg/mL), as well as Gram-negative *P. aeruginosa* (2.5 mg/mL) and *E. coli* (2.5 mg/mL), were the same for all the different extraction temperatures. The strongest antibacterial activity was observed towards *S. aureus* with an MIC of 0.3 mg/mL and an MBC of 0.6 mg/mL for the *G. hederacea* extract obtained at 40 °C/250 bar. Among Gram-negative bacteria, the strongest inhibition of bacterial growth was observed for *P. aeruginosa* for all tested extracts.

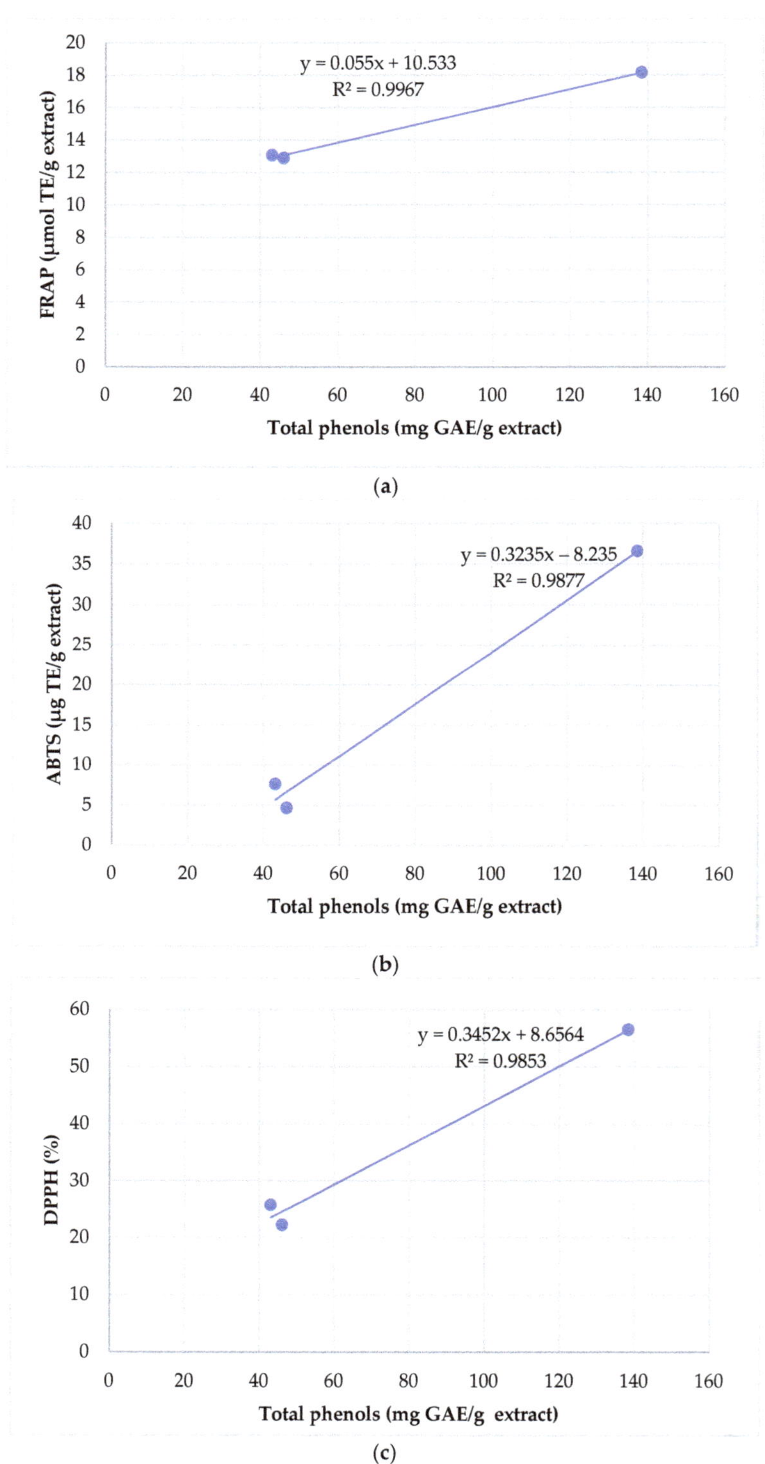

Figure 1. Correlation between TPC and (**a**) FRAP, (**b**) ABTS and (**c**) DPPH results of *G. hederacea* extracts.

Table 3. Antibacterial activity of *G. hederacea*, displayed by SC-CO_2 extracts, against tested indicator microorganisms.

Microorganism	MIC/ MBC/ (mg/mL)	*Glechoma hederacea* Extracts, Extraction Conditions:		
		40 °C	50 °C	60 °C
Gram-positive bacteria				
S. aureus ATCC 33862	MIC	0.3	2.5	2.5
	MBC	0.6	>5.0	2.5
B. subtilis ATCC 11774	MIC	0.6	5.0	>5.0
	MBC	1.25	>5.0	>5.0
E. faecalis ATCC 19433	MIC	0.6	2.5	2.5
	MBC	1.25	5.0	2.5
Micrococcus luteus ATCC 4698	MIC	1.25	1.25	1.25
	MBC	5.0	>5.0	>5.0
Gram-negative bacteria				
P. aeruginosa ATCC 9027	MIC	1.25	1.25	1.25
	MBC	2.5	5.0	2.5
S. Enteritidis ATCC 13076	MIC	1.25	2.5	2.5
	MBC	2.5	5.0	2.5
E. coli ATCC 8739	MIC	2.5	2.5	2.5
	MBC	5.0	5.0	2.5

The antibacterial potential of *G. hederacea* extracts obtained by SC-CO_2, depending on the process conditions, has not been established in the literature; however, the influence of different extraction conditions of SC-CO_2 on the antibacterial activity of plant extracts is well described. Cadena-Carrera et al. (2019) [60] studied the biological activity of SC-CO_2 guayusa leaf (*Ilex guayusa* Loes.) extracts obtained with differentiated process conditions. The authors tested, among others, antibacterial properties against *B. subtilis*, *S. aureus*, *E. coli* and *P. aeruginosa*. On the contrary to the results obtained in the studies with guayusa leaves, the lower extraction temperature of *G. hederacea* extracts resulted in higher antibacterial properties for most of the bacterial strains. The obtained results indicate the influence of the different extraction temperatures of SC-CO_2 on the total polyphenol composition of *G. hederacea* extracts and consequently on the antibacterial activity. Mendiola et al. (2008) [61] determined antimicrobial activity of SC-CO_2 extracts of green alga (*Dunaliella salina*) obtained using different extraction parameters against *E. coli*, *S. aureus*, *C. albicans* and *Aspergillus niger*. The results showed that all tested alga extracts presented antimicrobial activity against selected bacteria and yeasts. The authors confirmed the findings that when comparing the activities of the extracts obtained under the different experimental conditions, the sample obtained at the lowest temperature (9.8 °C) was the most active. Therefore, it can be clearly stated that the lower the SC-CO_2 extraction temperature, the higher antibacterial properties the extract demonstrated. It has been reported that *G. hederacea* extracts show antimicrobial activity against some microorganisms; however, their antibacterial activity is related to the extraction techniques used, to the different parameters of the process and therefore the concentration of active substances, and finally with the tested microorganism strains [22,62,63].

3.4. Fungistatic Activity—MIC and MFC Values

The fungistatic activity of *G. hederacea* extracts obtained via SC-CO_2 against five filamentous fungi, expressed as MIC and MFC values, are presented in Table 4 and Figure S1a–e in the Supplementary Materials. Based on the results, it can be stated that the tested extracts exhibited fungistatic activity towards the tested fungi; however, this activity was dependent mainly on the fungal species, whereas the extraction conditions generally did not affect the activity of the extracts. An exception was the extract obtained at 40 °C, which demonstrated activity against *B. cinerea* (MIC and MFC values were 5 mg/mL), whereas extracts obtained at 50 °C and 60 °C showed no inhibition of this species in the tested concentration range. Similar results can be observed in the case of *C. albicans*, where the extract obtained at 40 °C demonstrated the highest antagonistic activity compared to the other tested extracts. The strongest activity of all extracts was observed towards *S. sclerotiorum* with MIC and MFC values of 1.25 mg/mL, as well as towards *C. albicans* (MIC and MFC at the level of 1.25 mg/mL), but only for the extract obtained at 40 °C. Among the *Fusarium* strains, higher sensitivity to the tested *G. hederacea* extracts was observed for *F. graminearum* (MIC and MFC at 2.5 mg/mL) compared to *F. culmorum* (MIC and MFC at 5.0 mg/mL level). No fungistatic activity of the tested extracts was observed against *A. alternata*.

Table 4. Fungistatic activity of *G. hederacea* extracts obtained via SC-CO_2 against tested indicator fungi.

Microorganism	MIC/ MFC (mg/mL)	*Glechoma hederacea* Extracts, Extraction Conditions:		
		40 °C	50 °C	60 °C
F. graminearum KZF 1	MIC	2.5	2.5	2.5
	MFC	2.5	2.5	2.5
F. culmorum KZF 5	MIC	5.0	5.0	5.0
	MFC	5.0	5.0	5.0
A. alternata KZF 13	MIC	5.0	5.0	5.0
	MFC	5.0	5.0	5.0
S. sclerotiorum KZF 23	MIC	1.25	1.25	1.25
	MFC	1.25	1.25	1.25
B. cinerea BPR 187	MIC	5.0	>5.0	>5.0
	MFC	5.0	>5.0	>5.0
C. albicans ATCC 10231	MIC	1.25	2.5	>5.0
	MFC	1.25	>5.0	>5.0

Interesting data about the diversified fungistatic activity of different plant extracts obtained via SC-CO_2 can be found in the literature. Many authors reported that the antagonistic activity of the extracts strongly depends on the fungal genus, which was also underlined in this study. Confortin et al. (2019) [64] showed no differences in the fungistatic activity of extracts obtained via SC-CO_2 from *Lupinus albescens* against *F. oxysporum* and *F. verticillioides*, which is similar to the result obtained in the present work (the same MIC and MFC values were obtained for *F. graminearum* and *F. culmorum*). The high dependence of the antifungal activity of extracts of guayusa leaves (*Ilex guayusa* Loes.) obtained by SC-CO_2 on the type of indicator microorganism was shown by Cadena-Carrera et al. (2019) [60]. The studies presented in the mentioned paper showed little impact (in the case of *E. floccosum*, *M. canis*, *M. gypseum* and *T. mentagrophytes*) or no influence (in the case of *A. fumigatus*, *Rhizopus* and *C. albicans*) of the extraction conditions tested on the fungistatic activity of the extracts obtained via SC-CO_2 from guayusa leaves [60]. In the presented study, a minor influence of extraction conditions on the fungistatic activity of *G. hederacea* extracts was also noted. However, in the case of *C. albicans*, a significantly stronger effect of the extract obtained at 40 °C was observed compared to the other temperature variants.

3.5. Antibiofilm Activity

The antibiofilm activity of the tested *G. hederacea* extracts, expressed as a percentage of inhibition of biofilm formation or biofilm removal, is presented in Figure 2. Four strains of bacteria were used for this experiment. All tested extracts reduced the biofilm formation, with the highest ratio demonstrated by extract obtained at the 40 °C compared to the other temperature variants. The percentage inhibition of the biofilm formation by *E. coli* and *E. faecalis* exceeded 90%, whereas in the case of *P. aeruginosa* and *B. subtilis* it was almost 90% (88.6% and 87.9%, respectively). The extract obtained at 50 °C also had a strong inhibitory effect on biofilm formation, displaying the strongest inhibition of biofilm formed by *E. coli* (88.2%). The percentage reduction of biofilm formation by other bacteria was lower, with the lowest percentage of biofilm inhibition by *B. subtilis* (55.0%). In the case of the extract obtained at 60 °C, the best results in the reduction of biofilm formation were observed for *E. coli* (80.7%) and *E. faecalis* (71.5%), whereas the inhibitory efficiency was lower for *B. subtilis* (60.5%) and *P. aeruginosa* (38.7%).

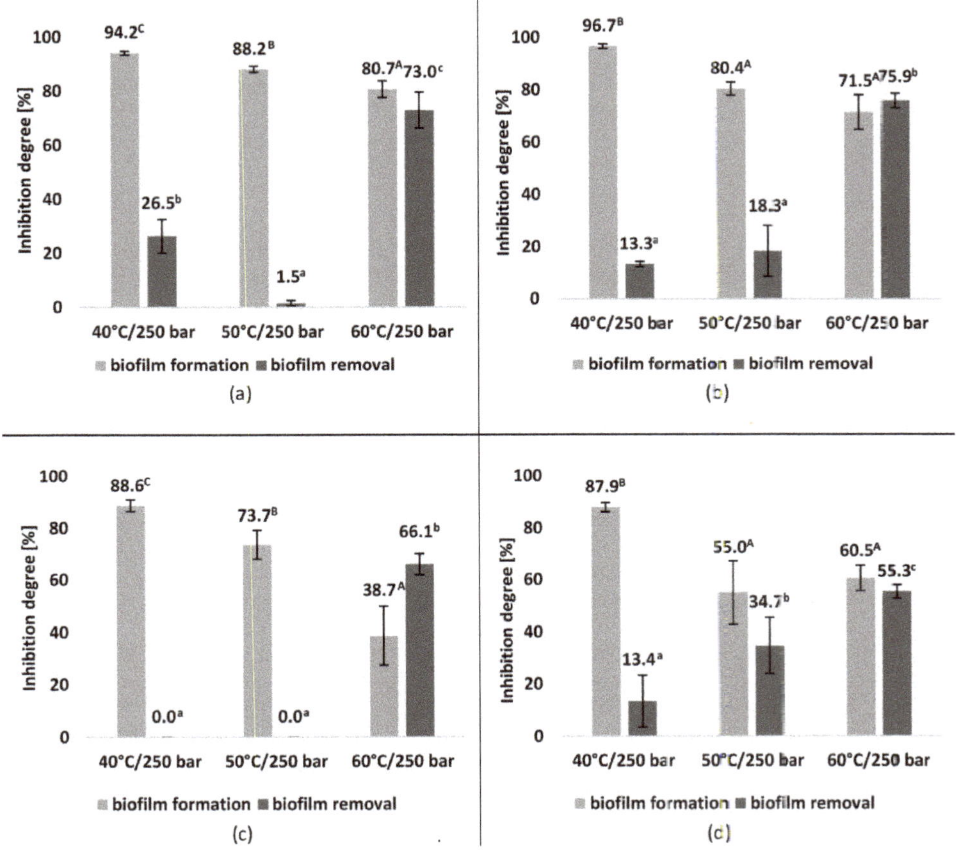

Figure 2. Effect of *G. hederacea* SC-CO_2 extracts on biofilm formation and removal: (**a**) *E coli*, (**b**) *E. faecalis*, (**c**) *P. aeruginosa*, (**d**) *B. subtilis*. Averages with different letters (A–C) for biofilm formation and (a–c) for biofilm removal are significantly different at $p < 0.05$.

The tested extracts showed variable effects on the removal of mature biofilm. In contrast to the inhibition of biofilm formation, the best results in biofilm removal were observed after the application of extracts obtained at 60 °C, within the range of 55.3–75.9%, depending on the microorganism. Only a small percentage of biofilm was removed by extracts obtained at 40 °C and 50 °C. Depending on the extract and bacterial strain, the efficiency of biofilm destruction ranged from 1.5% to 34.7%, with no effect on the biofilm formed by *P. aeruginosa*.

The antibiofilm activity of *G. hederacea* has not been described in the literature; however, several studies have described the effect of plant extracts obtained via supercritical fluid extraction. Al-Maqtari et al. (2020) [65] reported that extracts of *Artemisia arborescens*, *Artemisia abyssinica*, *Pulicaria jaubertii* and *Pulicaria petiolaris* were effective as anti-biofilm formation agents for all tested bacteria at 1/2 MIC. The highest inhibition rate of biofilm formation by the extracts was observed against *B. subtilis*, whereas the lowest inhibition ratio was noted on *K. pneumoniae*, *S. typhimurium*, *P. aeruginosa* and *E. faecalis*. Abdullah et al. (2021) [66] found that green cardamom essential oil obtained via supercritical fluid extraction prevented the biofilm formation of *E. coli* O157:H7 and *S. typhimurium* JSG 1748. In the presented work, the antibiofilm activity also depended on the bacterial strain, with the highest inhibition rate observed against *E. coli* and *E. faecalis*, whereas the biofilm prevention performance was slightly lower for *P. aeruginosa* and *B. subtilis*. Moreover, the extraction conditions had an impact on the degree of reduction of biofilm formation. In a study by Santos et al. (2021) [67], propolis extracts obtained via the supercritical and ethanolic extraction methods were effective in interfering with bacterial biofilm formation, whereas only little activity was observed on the consolidated film, which is consistent with the results obtained in the presented work. As the literature data suggest, one of the reasons for the extracts' weak efficacy in disrupting the biofilm maybe due to the structure of the biofilm, as the exopolymeric matrix may prevent the penetration of antibacterial agents [68].

4. Conclusions

The results of antioxidant, antimicrobial and antibiofilm activity analysis of *G. hederacea* extracts obtained via SC-CO_2 under different conditions of the process using methanol as a co-solvent show that the temperature of extraction affects the biological activity of the tested product. Overall, the results indicate that the SC-CO_2 extracts are characterized by high TPC values that differ depending on the extraction conditions. The high TPC values correlate with high antioxidant properties, as well as antimicrobial (excluding filamentous fungi) and antibiofilm activity (the prevention of biofilm formation). Among the tested extracts, the most promising results were obtained for the extract obtained at 40 °C, including the highest TPC value, as well as the best-performing antimicrobial and antioxidant properties. All tested extracts were effective in controlling biofilm formation of the studied bacteria at MIC concentration, but only the extract obtained at 60 °C efficiently removed formed biofilm. *G. hederacea* extracts obtained using SC-CO_2 could have promising applications at different stages of food production, as well as in the industry as a safe alternative to chemical preservatives or disinfectants due to their demonstrated properties.

Supplementary Materials: The following supporting information can be downloaded at: https://www.mdpi.com/article/10.3390/app12073572/s1, Figure S1: The fungistatic activity of obtained *G. hederacea* CO_2 extracts against (a) *F. graminearum*, (b) *F. culmorum*, (c) *A. alternata*, (d) *B. cinerea* and (e) *S. sclerotiorum*, expressed as MIC and MFC values. C—control; M—methanol; GH 40—*G. hederacea* CO_2 extract (extraction conditions: 40 °C/250 bar); GH 50—*G. hederacea* CO_2 extract (extraction conditions: 50 °C/250 bar); GH 60—*G. hederacea* CO_2 extract (extraction conditions: 60 °C/250 bar); Cp—percentage concentration of CO_2 extracts.

Author Contributions: Conceptualization, D.G., K.M., K.J. and A.W.; methodology, D.G., K.J., K.M., P.A.U., S.F. and A.W.; validation, K.J., S.F. and A.W.; formal analysis, K.M., D.G., R.G. and A.W.; investigation, D.G., K.J., P.A.U., S.F., K.M. and A.W.; resources, D.G., K.M., K.J., R.G. and A.W.; data curation, K.J. and A.W.; writing—original draft preparation, D.G., K.M., K.J., P.A.U., S.F., R.G. and A.W.; writing—review and editing, D.G., K.M. and K.J.; visualization, D.G., K.J., P.A.U., S.F. and K.M.; supervision, D.G. and A.W.; project administration, R.G. All authors have read and agreed to the published version of the manuscript.

Funding: This research was funded by Polish National Science Centre, grant number 2018/31/B/NZ9/03485.

Institutional Review Board Statement: Not applicable.

Informed Consent Statement: Not applicable.

Data Availability Statement: Publicly archived datasets are not available.

Conflicts of Interest: The authors declare no conflict of interest.

References

1. Uyttendaele, M.; Franz, E.; Schlüter, O. Food Safety, a Global Challenge. *Int. J. Environ. Res. Public Health* **2016**, *13*, 67. [CrossRef]
2. World Health Organization. *Estimates of the Global Burden of Foodborne Diseases: Food Borne Disease Burden Epidemiology Reference Group 2007–2015*; WHO: Geneva, Switzerland, 2015.
3. Soares, L.S.; Almeida, R.C.; Cerqueira, E.S.; Carvalho, J.S.; Nunes, I.L. Knowledge, attitudes and practices in food safety and the presence of coagulase-positive staphylococci on hands of food handlers in the schools of Camaçari, Brazil. *Food Control* **2012**, *27*, 206–213. [CrossRef]
4. Mostafa, A.A.; Al-Askar, A.A.; Almaary, K.S.; Dawoud, T.M.; Sholkamy, E.; Bakri, M.M. Antimicrobial activity of some plant extracts against bacterial strains causing food poisoning diseases. *Saudi J. Biol. Sci.* **2017**, *25*, 361–366. [CrossRef] [PubMed]
5. Pandey, A.; Singh, P. Antibacterial activity of *Syzygium aromaticum* (clove) with metal ion effect against food borne pathogens. *Asian J. Plant Sci.* **2011**, *1*, 69–80.
6. European Food Safety Authority (EFSA); European Centre for Disease Prevention and Control (ECDC). The European Union summary report on trends and sources of zoonoses, zoonotic agents and food-borne outbreaks in 2017. *EFSA J.* **2018**, *16*, e05500. [CrossRef]
7. Galiè, S.; García-Gutiérrez, C.; Miguélez, E.M.; Villar, C.J.; Lombó, F. Biofilms in the Food Industry: Health Aspects and Control Methods. *Front. Microbiol.* **2018**, *9*, 898. [CrossRef]
8. Chilaka, C.A.; De Boevre, M.; Atanda, O.O.; De Saeger, S. The Status of Fusarium Mycotoxins in Sub-Saharan Africa: A Review of Emerging Trends and Post-Harvest Mitigation Strategies towards Food Control. *Toxins* **2017**, *9*, 19. [CrossRef]
9. Wild, C.P.; Gong, Y.Y. Mycotoxins and human disease: A largely ignored global health issue. *Carcinogenesis* **2009**, *31*, 71–82. [CrossRef]
10. Zain, M.E. Impact of mycotoxins on humans and animals. *J. Saudi Chem. Soc.* **2011**, *15*, 129–144. [CrossRef]
11. Khaneghah, A.M.; Farhadi, A.; Nematollahi, A.; Vasseghian, Y.; Fakhri, Y. A systematic review and meta-analysis to investigate the concentration and prevalence of trichothecenes in the cereal-based food. *Trends Food Sci. Technol.* **2020**, *102*, 193–202. [CrossRef]
12. Sales, M.D.C.; Costa, H.B.; Fernandes, P.M.B.; Ventura, J.A.; Meira, D.D. Antifungal activity of plant extracts with potential to control plant pathogens in pineapple. *Asian Pac. J. Trop. Biomed.* **2016**, *6*, 26–31. [CrossRef]
13. Gnat, S.; Majer-Dziedzic, B.; Nowakiewicz, A.; Trościańczyk, A.; Ziółkowska, G.; Jesionek, W.; Choma, I.; Dziedzic, R.; Zięba, P. Antimicrobial activity of some plant extracts against bacterial pathogens isolated from faeces of red deer (*Cervus elaphus*). *Pol. J. Vet. Sci.* **2017**, *20*, 697–706. [CrossRef] [PubMed]
14. Gonelimali, F.D.; Lin, J.; Miao, W.; Xuan, J.; Charles, F.; Chen, M.; Hatab, S.R. Antimicrobial Properties and Mechanism of Action of Some Plant Extracts Against Food Pathogens and Spoilage Microorganisms. *Front. Microbiol.* **2018**, *9*, 1639. [CrossRef] [PubMed]
15. Manandhar, S.; Luitel, S.; Dahal, R.K. In Vitro Antimicrobial Activity of Some Medicinal Plants against Human Pathogenic Bacteria. *J. Trop. Med.* **2019**, *2019*, 1–5. [CrossRef] [PubMed]
16. Kim, J.; Song, S.; Lee, I.; Kim, Y.; Yoo, I.; Ryoo, I.; Bae, K. Anti-inflammatory activity of constituents from *Glechoma hederacea* var. *longituba*. *Bioorgan. Med. Chem. Lett.* **2011**, *21*, 3483–3487. [CrossRef]
17. An, H.-J.; Jeong, H.-J.; Um, J.-Y.; Kim, H.-M.; Hong, S.-H. *Glechoma hederacea* inhibits inflammatory mediator release in IFN-γ and LPS-stimulated mouse peritoneal macrophages. *J. Ethnopharmacol.* **2006**, *106*, 418–424. [CrossRef]
18. Chou, S.T.; Chan, Y.R.; Chung, Y.C. Studies on the Antimutagenicity and Antioxidant Activity of the Hot Water Extract of *Glechoma hederacea*. *J. Food Drug Anal.* **2012**, *20*, 637–645. [CrossRef]
19. Belščak-Cvitanović, A.; Durgo, K.; Bušić, A.; Franekić, J.; Komes, D. Phytochemical Attributes of Four Conventionally Extracted Medicinal Plants and Cytotoxic Evaluation of Their Extracts on Human Laryngeal Carcinoma (HEp2) Cells. *J. Med. Food* **2014**, *17*, 206–217. [CrossRef]
20. Chao, W.; Chan, W.; Ma, H.; Chou, S. Phenolic acids and flavonoids-rich *Glechoma hederacea* L. (Lamiaceae) water extract against H_2O_2—Induced apoptosis in PC12 cells. *J. Food Biochem.* **2021**, *46*, e14032. [CrossRef]
21. Cho, K.; Choi, Y.-J.; Ahn, Y.H. Identification of Polyphenol Glucuronide Conjugates in *Glechoma hederacea* var. *longituba* Hot Water Extracts by High-Performance Liquid Chromatography-Tandem Mass Spectrometry (HPLC-MS/MS). *Molecules* **2020**, *25*, 4713. [CrossRef]
22. Zhou, Y.; He, Y.-J.; Wang, Z.-J.; Hu, B.-Y.; Xie, T.-Z.; Xiao, X.; Zhou, Z.-S.; Sang, X.-Y.; Luo, X.-D. A review of plant characteristics, phytochemistry and bioactivities of the genus Glechoma. *J. Ethnopharmacol.* **2021**, *271*, 113830. [CrossRef] [PubMed]
23. Chou, S.-T.; Lin, T.-H.; Peng, H.-Y.; Chao, W.-W. Phytochemical profile of hot water extract of *Glechoma hederacea* and its antioxidant, and anti-inflammatory activities. *Life Sci.* **2019**, *231*, 116519. [CrossRef] [PubMed]
24. Kumarasamy, Y.; Cox, P.J.; Jaspars, M.; Nahar, L.; Sarker, S.D. Isolation, structure elucidation and biological activity of hederacine A and B, two unique alkaloids from *Glechoma hederaceae*. *Tetrahedron* **2003**, *59*, 6403–6407. [CrossRef]
25. Zhu, Q.-F.; Wang, Y.-Y.; Jiang, W.; Qu, H.-B. Three new norlignans from *Glechoma longituba*. *J. Asian Nat. Prod. Res.* **2013**, *15*, 258–264. [CrossRef]
26. Grabowska, K.; Żmudzki, P.; Wróbel-Biedrawa, D.; Podolak, I. Simultaneous Quantification of Ursolic and Oleanolic Acids in *Glechoma hederacea* and *Glechoma hirsuta* by UPLC/MS/MS. *Planta Medica* **2021**, *87*, 305–313. [CrossRef]
27. Chou, S.-T.; Lai, C.-C.; Lai, C.-P.; Chao, W.-W. Chemical composition, antioxidant, anti-melanogenic and anti-inflammatory activities of *Glechoma hederacea* (Lamiaceae) essential oil. *Ind. Crop. Prod.* **2018**, *122*, 675–685. [CrossRef]

28. Hwang, J.; Erkhembaatar, M.; Gu, D.; Lee, S.; Lee, C.; Shin, D.M.; Lee, Y.; Kim, M. *Glechoma hederacea* Suppresses RANKL-mediated Osteoclastogenesis. *J. Dent. Res.* **2014**, *93*, 685–690. [CrossRef]
29. Oalđe, M.; Kolarević, S.; Živković, J.; Aradski, A.A.; Marić, J.J.; Kolarević, M.K.; Đorđević, J.; Marin, P.D.; Šavikin, K.; Vuković-Gačić, B.; et al. A comprehensive assessment of the chemical composition, antioxidant, genoprotective and antigenotoxic activities of Lamiaceae species using different experimental models in vitro. *Food Funct.* **2021**, *12*, 3233–3245. [CrossRef]
30. Milovanovic, M.; Zivkovic, D.; Vucelic-Radovic, B. Antioxidant Effects of *Glechoma hederacea* as a Food Additive. *Nat. Prod. Commun.* **2010**, *5*. [CrossRef]
31. Chao, W.W.; Liou, Y.J.; Ma, H.T.; Chen, Y.H.; Chou, S. Phytochemical composition and bioactive effects of ethyl acetate fraction extract (EAFE) of *Glechoma hederacea* L. *J. Food Biochem.* **2021**, *45*, e13815. [CrossRef]
32. Varga, L.; Engel, R.; Szabo, K.; Abrankó, L.; Gosztola, B.; Németh, Z.; Sárosi, S. Seasonal Variation in Phenolic Content and Antioxidant Activity of *Glechoma hederacea* L. Harvested from Six Hungarian Populations. *Acta Aliment.* **2016**, *45*, 268–276. [CrossRef]
33. Benito-Román, O.; Rodríguez-Perrino, M.; Sanz, M.T.; Melgosa, R.; Beltrán, S. Supercritical carbon dioxide extraction of quinoa oil: Study of the influence of process parameters on the extraction yield and oil quality. *J. Supercrit. Fluids* **2018**, *139*, 62–71. [CrossRef]
34. Fornari, T.; Vicente, G.; Vázquez, E.; Garcia-Risco, M.R.; Reglero, G. Isolation of essential oil from different plants and herbs by supercritical fluid extraction. *J. Chromatogr. A* **2012**, *1250*, 34–48. [CrossRef] [PubMed]
35. Uwineza, P.A.; Waśkiewicz, A. Recent Advances in Supercritical Fluid Extraction of Natural Bioactive Compounds from Natural Plant Materials. *Molecules* **2020**, *25*, 3847. [CrossRef]
36. Bai, X.; Aimila, A.; Aidarhan, N.; Duan, X.; Maiwulanjiang, M. Chemical constituents and biological activities of essential oil from *Mentha longifolia*: Effects of different extraction methods. *Int. J. Food Prop.* **2020**, *23*, 1951–1960. [CrossRef]
37. Gonçalves, R.M.; Lemos, C.O.T.; Leal, I.C.R.; Nakamura, C.V.; Cortez, D.A.G.; Da Silva, E.A.; Cabral, V.F.; Cardozo-Filho, L. Comparing Conventional and Supercritical Extraction of (−)-Mammea A/BB and the Antioxidant Activity of *Calophyllum brasiliense* Extracts. *Molecules* **2013**, *18*, 6215–6229. [CrossRef]
38. Uwineza, P.A.; Gramza-Michałowska, A.; Bryła, M.; Waśkiewicz, A. Antioxidant Activity and Bioactive Compounds of *Lamium album* Flower Extracts Obtained by Supercritical Fluid Extraction. *Appl. Sci.* **2021**, *11*, 7419. [CrossRef]
39. Arabshahi-Delouee, S.; Urooj, A. Antioxidant properties of various solvent extracts of mulberry (*Morus indica* L.) leaves. *Food Chem.* **2007**, *102*, 1233–1240. [CrossRef]
40. Li, H.-B.; Cheng, K.-W.; Wong, C.-C.; Fan, K.-W.; Chen, F.; Jiang, Y. Evaluation of antioxidant capacity and total phenolic content of different fractions of selected microalgae. *Food Chem.* **2007**, *102*, 771–776. [CrossRef]
41. Mabrouki, H.; Duarte, C.; Akretche, D.E. Estimation of Total Phenolic Contents and In Vitro Antioxidant and Antimicrobial Activities of Various Solvent Extracts of *Melissa officinalis* L. *Arab. J. Sci. Eng.* **2017**, *43*, 3349–3357. [CrossRef]
42. Benzie, I.F.; Devaki, M. The ferric reducing/antioxidant power (FRAP) assay for non-enzymatic antioxidant capacity: Concepts, procedures, limitations and applications. *Meas. Antioxid. Act. Capacit.* **2018**, 77–106. [CrossRef]
43. Tomasina, F.; Carabio, C.; Celano, L.; Thomson, L. Analysis of two methods to evaluate antioxidants. *Biochem. Mol. Biol. Educ.* **2012**, *40*, 266–270. [CrossRef] [PubMed]
44. Re, R.; Pellegrini, N.; Proteggente, A.; Pannala, A.; Yang, M.; Rice-Evans, C. Antioxidant activity applying an improved ABTS radical cation decolorization assay. *Free Radic. Biol. Med.* **1999**, *26*, 1231–1237. [CrossRef]
45. Xiao, F.; Xu, T.; Lu, B.; Liu, R. Guidelines for antioxidant assays for food components. *Food Front.* **2020**, *1*, 60–69. [CrossRef]
46. Moradi, M.-T.; Karimi, A.; Alidadi, S.; Hashemi, L. In Vitro Anti-adenovirus Activity, Antioxidant Potential and total Phenolic Compounds of *Melissa officinalis* L. (Lemon Balm) Extract. *Int. J. Pharmacogn. Phytochem. Res.* **2016**, *8*, 1471–1477.
47. Gwiazdowski, R.; Gwiazdowska, D.; Marchwińska, K.; Juś, K.; Szutowska, J.; Bednarek-Bartsch, A.; Danielewicz, B. Fungistatic activity of essential oils towards selected oilseed rape pathogens. *Prog. Plant Prot.* **2018**, *58*, 300–305. [CrossRef]
48. Rzemieniecki, T.; Gwiazdowska, D.; Rybak, K.; Materna, K.; Juś, K.; Pernak, J. Synthesis, Properties, and Antimicrobial Activity of 1-Alkyl-4-hydroxy-1-methylpiperidinium Ionic Liquids with Mandelate Anion. *ACS Sustain. Chem. Eng.* **2019**, *7*, 15053–15063. [CrossRef]
49. Somrani, M.; Inglés, M.-C.; Debbabi, H.; Abidi, F.; Palop, A. Garlic, Onion, and Cinnamon Essential Oil Anti-Biofilms' Effect against *Listeria monocytogenes*. *Foods* **2020**, *9*, 567. [CrossRef]
50. O'Toole, G.A. Microtiter Dish Biofilm Formation Assay. *J. Vis. Exp.* **2011**, *47*, e2437. [CrossRef]
51. Silva, S.G.; de Oliveira, M.S.; Cruz, J.N.; da Costa, W.A.; da Silva, S.H.M.; Maia, A.A.B.; de Sousa, R.L.; Junior, R.N.C.; Andrade, E.H.D.A. Supercritical CO_2 extraction to obtain *Lippia thymoides* Mart. & Schauer (Verbenaceae) essential oil rich in thymol and evaluation of its antimicrobial activity. *J. Supercrit. Fluids* **2020**, *168*, 105064. [CrossRef]
52. Bezerra, F.; de Oliveira, M.; Bezerra, P.; Cunha, V.; Silva, M.; da Costa, W.; Pinto, R.; Cordeiro, R.; da Cruz, J.; Neto, A.C.; et al. Extraction of bioactive compounds. In *Green Sustainable Process for Chemical and Environmental Engineering and Science*; Elsevier: Amsterdam, The Netherlands, 2020; pp. 149–167.
53. Hahm, Y.H.; Cho, K.; Ahn, Y.H. Compositional Characteristics of Glucuronide Conjugates in Regional *Glechoma hederacea* var. *longituba* Herbal Extracts Using a Set of Polyphenolic Marker Compounds. *Plants* **2021**, *10*, 2353. [CrossRef] [PubMed]
54. De Melo, M.M.R.; Silvestre, A.J.D.; Silva, C.M. Supercritical fluid extraction of vegetable matrices: Applications, trends and future perspectives of a convincing green technology. *J. Supercrit. Fluids* **2014**, *92*, 115–176. [CrossRef]

55. Feliciano, R.P.; Meudt, J.J.; Shanmuganayagam, D.; Metzger, B.T.; Krueger, C.G.; Reed, J.D. Supercritical Fluid Extraction (SFE) of Cranberries Does Not Extract Oligomeric Proanthocyanidins (PAC) but Does Alter the Chromatography and Bioactivity of PAC Fractions Extracted from SFE Residues. *J. Agric. Food Chem.* **2014**, *62*, 7730–7737. [CrossRef]
56. García-Abarrio, S.; Martin, L.; Burillo, J.; Della Porta, G.; Mainar, A. Supercritical fluid extraction of volatile oil from *Lippia alba* (Mill.) cultivated in Aragón (Spain). *J. Supercrit. Fluids* **2014**, *94*, 206–211. [CrossRef]
57. Akowuah, G.; Mariam, A.; Chin, J. The effect of extraction temperature on total phenols and antioxidant activity of *Gynura procumbens* leaf. *Pharmacogn. Mag.* **2009**, *5*, 81.
58. Shahidi, F.; Zhong, Y. Measurement of antioxidant activity. *J. Funct. Foods* **2015**, *18*, 757–781. [CrossRef]
59. Matkowski, A. Antioxidant activity of extracts and different solvent fractions of *Glechoma hederacea* L. and *Orthosiphon stamineus* (Benth.) Kudo. *Adv. Clin. Exp. Med.* **2008**, *17*, 615–624.
60. Cadena-Carrera, S.; Tramontin, D.P.; Cruz, A.B.; Cruz, R.C.B.; Müller, J.M.; Hense, H. Biological activity of extracts from guayusa leaves (*Ilex guayusa* Loes.) obtained by supercritical CO_2 and ethanol as cosolvent. *J. Supercrit. Fluids* **2019**, *152*, 104543. [CrossRef]
61. Mendiola, J.A.; Santoyo, S.; Cifuentes, A.; Reglero, G.; Ibáñez, E.; Señoráns, F.J. Antimicrobial Activity of Sub- and Supercritical CO_2 Extracts of the Green Alga *Dunaliella salina*. *J. Food Prot.* **2008**, *71*, 2138–2143. [CrossRef]
62. Tian, F.M.; Huang, Z.H.; Wang, H.; Chen, Q.; Zhuo, P.Q.; Chen, W.D. The study of antibacterial activity of *Glechoma* Alcohol extraction in vitro. *J. Gansu Norm. Coll.* **2016**, *21*, 42–45.
63. Coss, E.; Kealey, C.; Brady, D.; Walsh, P. A laboratory investigation of the antimicrobial activity of a selection of western phytomedicinal tinctures. *Eur. J. Integr. Med.* **2018**, *19*, 80–83. [CrossRef]
64. Confortin, T.C.; Todero, I.; Soares, J.F.; Luft, L.; Brun, T.; Rabuske, J.E.; Nogueira, C.U.; Mazutti, M.A.; Zabot, G.L.; Tres, M.V. Extracts from *Lupinus albescens*: Antioxidant power and antifungal activity in vitro against phytopathogenic fungi. *Environ. Technol.* **2018**, *40*, 1668–1675. [CrossRef] [PubMed]
65. Al-Maqtari, Q.A.; Al-Ansi, W.; Mahdi, A.A.; Al-Gheethi, A.A.S.; Mushtaq, B.S.; Al-Adeeb, A.; Wei, M.; Yao, W. Supercritical fluid extraction of four aromatic herbs and assessment of the volatile compositions, bioactive compounds, antibacterial, and anti-biofilm activity. *Environ. Sci. Pollut. Res.* **2021**, *28*, 25479–25492. [CrossRef] [PubMed]
66. Abdullah; Asghar, A.; Algburi, A.; Huang, Q.; Ahmad, T.; Zhong, H.; Javed, H.U.; Ermakov, A.M.; Chikindas, M.L. Anti-biofilm Potential of *Elletaria cardamomum* Essential Oil against *Escherichia coli* O157:H7 and *Salmonella* Typhimurium JSG 1748. *Front. Microbiol.* **2021**, *12*, 620227. [CrossRef] [PubMed]
67. Santos, L.M.; Rodrigues, D.M.; Kalil, M.A.; Azevedo, V.; Meyer, R.; Umsza-Guez, M.A.; Machado, B.A.; Seyffert, N.; Portela, R.W. Activity of Ethanolic and Supercritical Propolis Extracts in *Corynebacterium pseudotuberculosis* and Its Associated Biofilm. *Front. Vet. Sci.* **2021**, *8*, 965. [CrossRef]
68. Davenport, E.K.; Call, D.R.; Beyenal, H. Differential Protection from Tobramycin by Extracellular Polymeric Substances from *Acinetobacter baumannii* and *Staphylococcus aureus* Biofilms. *Antimicrob. Agents Chemother.* **2014**, *58*, 4755–4761. [CrossRef] [PubMed]

MDPI
St. Alban-Anlage 66
4052 Basel
Switzerland
Tel. +41 61 683 77 34
Fax +41 61 302 89 18
www.mdpi.com

Applied Sciences Editorial Office
E-mail: applsci@mdpi.com
www.mdpi.com/journal/applsci

www.ingramcontent.com/pod-product-compliance
Lightning Source LLC
LaVergne TN
LVHW070635170726
843515LV00004B/255

* 9 7 8 3 0 3 6 5 4 4 2 5 0 *